BOTANICAL CULTURE OF
Mughal India
(AD 1526–1707)

Versha Gupta

PARTRIDGE

Copyright © 2018 by Versha Gupta.

ISBN: Softcover   978-1-5437-0335-1
           eBook   978-1-5437-0336-8

All rights reserved. No part of this book may be used or reproduced by any means, graphic, electronic, or mechanical, including photocopying, recording, taping or by any information storage retrieval system without the written permission of the author except in the case of brief quotations embodied in critical articles and reviews.

Because of the dynamic nature of the Internet, any web addresses or links contained in this book may have changed since publication and may no longer be valid. The views expressed in this work are solely those of the author and do not necessarily reflect the views of the publisher, and the publisher hereby disclaims any responsibility for them.

Print information available on the last page.

To order additional copies of this book, contact
Partridge India
1663 Liberty Drive
Bloomington, IN 47403
000 800 10062 62
orders.india@partridgepublishing.com

www.partridgepublishing.com/india

CONTENTS

Preface xi
Introduction... xiii

Development of Horticulture 1
Royal Gardens 52
Society and Plantation of Trees 115
Botanical Products and Their Significance 140
Preservation of Fruits and Flowers 190

Appendix - I 211
Appendix - II 215
Appendix - III 217
Appendix - IV 221
Appendix - V 225
Bibliography 229
Index 245

MAPS

1. Areas famous for different varieties of fruits
2. Areas famous for tree products
3. Different Industries related to the Botanical Products

LIST OF ILLUSTRATIONS

1. CHASHMA SHAHI KASHMIR
2. NISHAT GARDEN KASHMIR
3. NASEEM BAGH SRINAGAR
4. PINJAUR GARDEN
5. WATER AS DECORATIVE FEATURE IN PINJAUR GARDEN
6. CHINAR TREE IN KASHMIR
7. BARADARI AT SHALAMAR GARDEN KASHMIR

PREFACE

Indian society since times immemorial has played an important role in the preservation and sustenance of the environment. Keeping this aspect in mind various environment friendly practices were either initiated afresh or were continued albeit with some modifications by the Indian society and enlightened rulers who had public welfare as their supreme motto. This trend of preserving environment and its surroundings was continued by the Mughal rulers with their notable contributions towards the initiation and propagation of botanical culture in india. The present work introduces the readers the culture of environmental protection which had been initiated and sustained, starting form the ancient and traversing through the Sultanate and the Mughal period. It minutely details the initiatives undertaken for the development of horticulture during the Mughal period. The work enumerates the contribution of the Mughal kings and nobility towards laying out the gardens on a exquisite scale. It also focuses on the activities initiated by the general public for the preservation of the ecology in the geographical areas inhabited by them. Various botanical products and the scientific inventions made in this field find due mention regarding their role in the upkeep of the economy and general prosperity of the society. The notable role of the religious elements of various hues and institutions established by them are the highlights of this work.

The historical analysis of this work is based on both the contemporary sources and modern works. The contemporary sources consist of the chronicles, archaeological remains, Mughal documents (*farmans* and *prawanas*) and travel accounts. The modern works pertaining to the social and economic history of medieval India are also utilised. The present study is an attempt to add botanical culture as a dominant part of the socio-economic life of the Mughal India. A survey of some existing Mughal gardens, particularly of the Punjab, Haryana and Kashmir has been made. Some of the photographs of these gardens and trees are appended in this work.

I express my sincere gratitude to the staffs of libraries of ICHR, Jawahar Lal Nehru University, Jamia Milia Islamia, Krishi Bhawan, the National Archives of India, New Delhi; Osmania University, Hyderabad; Bhuri Singh Museum, Chamba; Library of the Department of History, The Central Library, University of Jammu; Amar Palace library, the State Archives of Jammu and Srinagar, the Ranbir Library, Jammu for extending their help to me during my visits for the material collection.

I express my heartfelt and sincere gratitude to Professor Shailendra Singh Jamwal of Department of History, University of Jammu who always encouraged me in completing this task. I am greatly thankful to my teacher Professor Jigar Mohammad of Department of History, University of Jammu in helping me to bring out this work. I express my deep gratitude to Pranvee Fathepuri, Sana and Neeraj Jandial who helped me in accomplishing this task. I am also thankful to my publishing house Partridge India.

My thanks are due to the members of my family especially to my father, late shri Chander Prakash Gupta ji for encouraging me for acquiring higher studies and was happy to see me at academic work. I am also grateful to my mother Smt. Kamla Gupta for providing me material and moral assistance when it was needed. I take this opportunity to thank my husband Pawan Gupta and my son Siddharth who always assisted me and were source of inspiration in times of need and it is to my son this book is lovingly dedicated.

Versha Gupta.

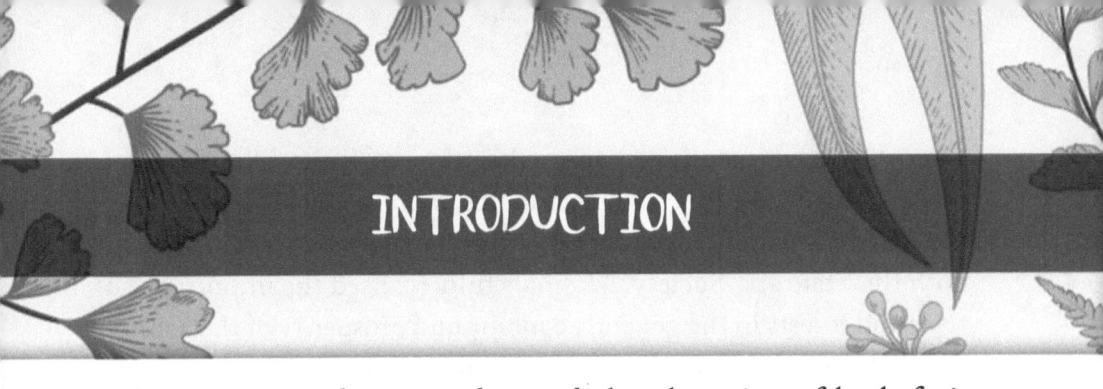

INTRODUCTION

Indian society and state understood the plantation of both fruit bearing and non-fruit bearing trees as an integral part of their socio-economic activities. Though agricultural productions were the dominant source of the livelihood of the society and source of state's revenue but right from the beginning people treated forest produce as an important source for fulfilling their day to day needs. Even before the man developed agriculture, he was dependent on trees for fulfilling his needs, he lived mostly on fruits and flesh. Man sought shelter of trees against inclement weather, got fruits and nuts to eat and wood for implements. It is from wood that he obtained fire which enabled him to cook his food and to warm his dwellings. And with the development of civilization the need for trees and tree products increased. Consequently, the culture of tree protection emerged in ancient Indian society.

The trend of tree plantation and protection of forests may be traced from ancient India. The plantation and protection of trees was understood as a pious work in the Indian society. During Harappan and *Vedic* ages certain trees were regarded as sacred notably the *pipal*. Even a seal of *Harappan* period depicts a horned goddess on a *pipal* tree. Aryans of the *Vedic* Age had great love for trees and flowers. In *Mahabharata* we find a numerous references of several trees such as *mango, asoka, champaka, nag-champa, sal,* coral *and* oleander, etc..

Continuity of cultural traditions in India is proverbial. This continuity is found in almost all the aspects of life including horticulture. During Sultanate period [1206 – 1526], the Sultans of Delhi worked for the improvement of Indian fruits and system of gardening as a whole. Firoz Tugluq (1351–1388) laid out 1200 gardens in the vicinity of Delhi. These gardens were embellished with the choicest fruits. Zia-ud-din Barni, Ibn-Batuta and Shams

Siraj Afif have mentioned the cultivation of fruits like mangoes, melons, sugarcane, limes, pomegranates, *narang*, etc. However, under the Mughals botanical products became a major source of income for both the State and Society. Mughals fully realised the importance of trees and forests in the general economy and prosperity of the country and followed a definite policy for their conservation and scientific exploitation. A large number of changes took place in the product and practices of horticulture. All the Mughal emperors from Akbar to Aurangzeb shared a natural gift of the traditional Muslim interest in horticultural pursuits and were keenly desirous of promoting this particular branch of activity. Many contemporary Mughal sources refer to the development of horticulture during this period. Abul Fazl, the court historian of the Mughal emperor Akbar(1546-1605) writes that Akbar invited the horticulturists from Iran and Turan and got them to settle down in India. Seeds were imported from the far off countries to grow the best quality of fruit in the country. With the introduction of grafting system, quality of various fruits were improved.

A large number of gardens were constructed throughout the empire. These gardens gave a fillip to horticulture and plantation of numerous trees as well as many herbs and shrubs. Experiments on grafting and plantation of new varieties of fruits were carried out in these gardens. The variety of plants found in Mughal gardens was great and by some accounts, there were so many different types that gardens could be kept continuous bloom all the year. A large number of new fruits such as papaya, cashew nut, pine-apple were introduced during this period. Mughal Emperors themselves participated in the promotion of horticulture. Cutting of trees was regarded as punishable act. Jahangir(1605-1627) himself writes in his memoirs that he got the thumbs of a man to be cut because he had destroyed the *Champa* tree. No revenue was collected from the orchards. Akbar remitted the cess on the fruit trees i.e. *Sar-i-darakhti*. Jahangir too exempted the orchard land from the taxes.

Not only the ruling class but common people also took great interest in plantation of trees as trees were associated with all the socio-economic activities of the people and furnished them with all their needs. The trees contributed to the development of both the rural and urban economy of the empire. People worked for the expansion of the tree plantation and avoided their destruction. Trees were also worshipped especially by the Hindus. In fact tree worship was possibly the earliest and the most prevalent form of religion in India. Not only the trees but their products were also regarded as sacred. The tree products such as fruits, flowers, sandal, betel leaf, etc. were used for the religious purposes. Tree products were also an important requirement in the social occasions. With the progress of horticulture and process of intensification, the industry of flowers and fruit preservation became one of the dominant features of the economy of the Mughal Empire. Fruits and flowers were preserved in various forms such as pickles, marmalades, scents, perfumes, etc. During the 16th and 17th centuries with the the development of the urban centres the need for tree products increased. These were related with almost every important activity, wether it was the construction of the buildings, manufacture of the means of transport, for medicinal purposes, maintenance of silk industry and the manufacture of the house hold articles. Trees not only fulfilled the needs of the people but also gave promotion to number of industries. These industries in turn provided employment to a number of artisans as they were engaged in the manufacture of the different products and their manufactures were not only appreciated in the country but also in the foreign lands. There was a brisk trade in these products. all this widened the scope of Indian goods in the foreign markets and helped to improve the economy of the Mughal empire.

    The value of this work lies in the fact that although many works on the economic activities such as agricultural productions, land revenue system and trade and commerce are produced by the modern historians but no detailed study of the botanical culture and horticulture in Mughal India is available. Therefore this work concentrates on the various aspects of the horticulture and botanical culture in India under the Mughals. Both the social and the economic significance of the botanical products have been covered. Role of the royal gardens in the expansion of horticulture and in the promotion of tree plantation in the different parts of the Mughal empire have been examined. The work is valuable for the study of the social and economic life during the Mughal period. From social point of view we find that there were many curious customs, traditions and superstitions among the people relating to trees. The work also concentrates upon the economic significance of the botanical products. The study provides a detailed account of the numerous tree products and their multiple uses. For instance, from Abul Fazl's account we find that about seventy-two types of wood was available in india and different varieties of wood was used in the building industry and for the manufacture of the means of transport. The botanical products led to the development of the numerous industries and provided employment to a large number of artisans. For instance, in building industry we come across a number of artisans such as carpenters, *pinjara saz, arrakash, patal band, lakhira*, etc. Both horticulture and botanical products gave boost to the economy of the Mughal empire. The income from the fruits from the orchards and gardens is recorded by the sources of the Mughal period which shows that the state was conscious of the economic value of the botanical products.

This work is also of considerable value to the environmentalists because it throws ample light on the significance of the different varieties of trees and their products especially in today's world where we are confronting the serious problems like global warming and climatic change which has occurred primarily because of the cutting of the trees and deforestation. In fact this work is important for every one because it makes us realise the importance of the trees in our life. More over this work also have some importance for the botanists because it provides along list of the numerous trees, plants, flowers, herbs and shrubs with their scientific names which were grown in India right from the ancient times.

DEVELOPMENT OF HORTICULTURE

Tree products were the major source of livelihood of the people of India from ancient period onwards. The trend of tree plantation and protection of forests may be traced from ancient India. The Aryans of the Vedic times were the great lovers of nature. They had great love for trees and flowers, and the very name they gave to flowers, *Sumanasa* – that which pleases the mind – reveals their aesthetic sensibility[1]. The ancient Indians greatly loved flowers and trees, specially the flowering trees, which were very frequently mentioned in poetry, especially the *asoka* (saraca Indica) – a smallish tree bearing a mass of lovely scarlet or orange blossoms; other favourites were the tall pale – flowered *sirisa* (Albizzia spp.), the fragrant, orange flowered *kadamba* (Anthocephalus cadamba*)*, and the *red Kimsuka* (Butea frondosa*)*; the banana *(kadali)* were grown for ornament as well as for its fruit. Bushes and creepers were also much loved especially the jasmine, of which there were many varieties; and the white *atimukta* (hiptage madablota); other popular trees were the *campka (michelia champaca)*, and the hibiscus, or the china rose (*japa*).[2] The *rishis* admired trees of forests because they were hardy and could survive the droughts. *Sal* trees in the month of June when grass dries up and the earth is brown look marvellously fresh and green.

> *Plant the mango, plant the tamarind, and plantain ;*
> *cluster of fruits will weigh their boughs*
> *plant ten kachnar trees for flowers;*
> *In a garden set the tulsi*
> *water them unweariedly, but they will wither*
> *but trees in the forests,*
> *which depend upon God alone,*
> *never wither and die*
> *the forest trees grow always*[3]

Hiuen Tsang, the famous Chinese pilgrim in India (A.D. 630) has given a list of fruits which he saw in India. The main fruits mentioned by him are "the *amala* fruit (*ngan-mi-lo*), the *maduka* fruit (*mo-tu-kia*), the *bhadra* fruit (*pa-ta-lo*), the *kapittha* fruit (*kie-pi-ta*),... the *mocha* fruit (*mau-che*), the *narikela* fruit (*na-li-ki-lo*), the *panasa* fruit (*pan-na-so*)... as for the date (*tasu*), the chestnut (*lih*), the loquat (*p'l*), and the persimmon (*thi*) they are not known. The pear (*li*), the wild plum (*nai*), the peach (*t'au*), the apricot (*hang* or *mui*), the grape (*po-tau*), etc., these all have been brought from the country of Kashmir and are found growing on every side. Pomegranates and sweet oranges are grown everywhere".[3a.] The kings in ancient India made great endeavours to plant trees for instance Asoka (264 – 227 B.C.) encouraged the planting of trees in gardens and along the roads in form of avenues.[4] The tree products also contributed to the economy of the state in the ancient India and its is evident from the laws of Manu, which specify the king's taxes in the subjoined terms *He may also take a 6th part of the clear annual increase of trees, ... medicinal substances, perfumes, liquids, flowers, roots and fruits...*[5]

Continuity of cultural traditions in India is proverbial. This continuity is found in almost all the aspects of life including horticulture. Religious injunctions both in Hinduism and in Islam, helped in planting trees all over the land; trees that were helpful to man in the economic sense of the word.[6] During the Sultanate period (1206–1526), the sultans of Delhi worked for the improvement of Indian fruits and system of gardening as a whole. From Alauddin Khalji (1295–1315) to Akbar (1556–1605) there was an emphasis on the planting trees that were useful as well as charming from aesthetic point of view. Firoz Tugluq (1351–1388) was exceedingly fond of laying out gardens and is credited with the laying of 1200 gardens in the vicinity of Delhi, these gardens were embellished with the choicest fruits. These efforts led to a general improvement in the quality of the most of fruits. In every garden there were white and black grapes and fruits of rich variety. The grapes became so cheap that they sold at one *jital* per *seer*. This contributed to state economy also. The state's share of income from these gardens amounted to

80,000 *tankas* annually.[7] Moreover the canals of Firoz Shah which supplied water to the new cities of Hisar Firoza and Firozabad, they gave a great flip to the horticulture of the areas. Historian Ziauddin Barni has visualised the cultivation of fruits like mangoes, melons, figs, oranges, limes, pomegranates and apples and flowers like roses, etc.[8] Other fruits mentioned by Historian Sirajuddin Afif, a contemporary of Firoz Shah Tuglaq are:

Sugarcane, both black and ponda; sadaphal, jundhari, narang and sikandwal, etc.[9]

Father Monserrate has mentioned that Firoz Tugluq got trees to be planted in long avenues on both sides of the roads in his dominions. He writes:

Peruzis[10] *who was Pathan by race, was much devoted to piety; for he gave orders that through out his dominions, at the intervals of every two miles, resting places should be built in which a shady tree should be planted... He also planted trees in a long avenue on both sides of the roads, where there was room, in order that tired way fares might find shelter.*[11]

We get references to plantation of pleasant grooves of trees in Kashmir by Sultan Zain-ul Abidin (1420-1470). He built a palace in the middle of lake Ulur, and planted grooves of trees, so that there can be but few more agreeable places in the world.[12] Sikander Lodi (1488–1517), paid a special attention to the culture of pomegranates in Jodhpur, and he confidently declared that Persia could not produce pomegranates which would compare favourably with the Jodhpur variety in flavour.[13] Ibn Batuta in his work *Rehla* has described the fruits and trees of Hind: these fruits were mango *(amba)*, *kathal* (jack fruit), *tendu* (diospyros peregrine), *jambol (jaman, syzgiumcumini)*, sweet orange *(naranj)*, madhuka latifolia *(mahwa)*, *kasera* (scirpusgrossus) *and* pomegranate *(rumman)*.[14] The culture of flowers is of very ancient date in Hindustan. They have been remarkable for their charm and beauty e.g. marigold. Whole chapters of their books have been devoted by Amir Khusrau and Malik Muhammad Jaisi to the description of flowers of the land.[15]

During Mughal period horticulture received an special attention of the ruling class. A large number of changes took place in the product and practices of Horticulture. All the Mughal emperors from Babur (1526 –1530) to Aurangzeb (1658–1707) shared the natural gift of traditional Muslim interest in the horticultural pursuits and were keenly desirous of promoting and popularising this particular branch of activity. As one finds that a striking feature of India is the extraordinary variety of its physical aspects, vegetation and her people. Babur the first Mughal emperor had a keen sense of observation and his memoirs provide a vivid account of the country in the early sixteenth century. Coming from the temperate Central Asia, He felt the uniqueness of the country.

Hindustan is a remarkably fine country. Its hills and rivers, its forests and plains, its animals and plants, its inhabitants and their languages, its winds and rains are all of different nature.[15a]

Babur's love for fruits and natural beauty imbibed him with a desire for planning for gardens. On the banks of Jamuna, he raised several gardens with beautiful flowers and fruit trees.[16] His daughter Gulbadan Begam has described him as planter of trees.[17] Babur has given a detailed description of the fruits of Kabul and Samarkand. About the fruits of Samarkand he writes;

Grapes, melons, apples and pomegranates, all fruits indeed, are good in Samarqand; two are famous, its apples and its sahibi (grapes).[18]

Babur had so much love for fruits that when he was in India, he wept at the sight of the fruits that were sent to him from the Kabul.[19] Like Babur his successors also had great interest in the development of horticulture. Akbar (1556-1605) his grandson, a wise and tolerant statesman and a brilliant General, could spare time to set out plantations with his own hands. He had obtained Irani and Turani horticulturists and caused them to settle down in India. According to his court historian Abul Fazl:

His Majesty looks upon the fruits as one of the greatest gifts of the creator, and pays much attention to them. The horticulturists of Iran and Turan have, therefore, settled here, and the cultivation of trees is in a flourishing state.[20]

Akbar's son Jahangir (1605–1627) was even more devoted to his gardens and intensely interested in botany, observing fruits and flowers he saw upon his marches with meticulous care. He was very much fond of fruits especially mangoes, and had fruits served as relish with his drinks. He even exempted the cultivation of fruits from tax to encourage horticulture.[21] Shahjahan (1627–1658) often left the fortress (palace) early in the morning, and as a recreation would gather fruits in the garden in the company of his favourite pages.[22] Even during the reigns of the later Mughals period we get numerous references to the development of Horticulture from various travel accounts and contemporary Persian and vernacular sources. Bernier, a French Physician who lived in India from 1656 – 1668, has given the account of the fruit market of Delhi in which various types of good quality fruits were available. According to him pears and apples of all sorts were available and there were melons which lasted the whole winter.[23] Manucci, a Venetian has also praised the quality of the Indian fruits. He has also mentioned that trees were planted in series throughout the empire.[24] Gulam Hussain Zaidpuri has regarded mango as the best fruit of land.[25] In later period many provincial rulers took out the task of the extension of horticulture and improvement of the quality of fruits, for instance, Nawab Shuja-ud-Daula (1754-1765) of Awadh planted many mango trees and many other fruit trees at Faizabad.[26]

The Mughals sources are full of the description of trees and their products. Babur had a keen sense of observation and he mentions in his memoirs the plants, fruits, flowers that he saw in India. He was not much impressed by the quality of Indian fruits when he writes:

Hindustan is a country of few charms. Its people have no good looks... there are no good horses, no good dogs, no grapes, musk melons or first rate fruits...[27]

But it was not true of all the fruits as he has regarded some of the fruits such as mangoes, jack fruit, lote fruit, *karunda*, *Sangtara* as the first rate fruits.[28] Babur had so much love for mangoes that before the conquest of Hindustan, he prayed the God for the productions of Hind that is the mangoes. He prayed,

O God! if the government of Hindustan is destined to be given to me and mine, let these productions of Hind be brought presently before me, betel leaves and mangoes, and I shall accept them as omen.[29]

Babur's detailed description of the flora and fauna is fascinating, in some cases amusing. While he loved mangoes and plantains, but compared jack fruit to the revolting intestines of sheep. Babur's record of fruits and ornamental plant is important in the sense that we know in an authentic manner what plants and fruits were grown in India in early 16th Century. The fruits of India mentioned by Babur are mango, plantains, *ambli* (the tamarind), *mahua*, mimusops, *jaman*, *kamrak*, jack fruit, monkey jack, lote fruit, *karunda*, *paniyala*, *gular*, *amla*, date palm, coconut palm, *sangtara*, etc. Babur has given a detailed description of the India fruits and has described the manner in which, these fruits were eaten, the season during which these fruits were grown, their uses etc.[30] Babur has regarded mango (*ambah*), the best fruit of India. He writes that some people called it *anbah* and some called it *naghzak*. He mentions the verses of Khwaja Amir Khusrau, a noted historian and poet of the Delhi Sultanate, in which he has regarded mango as the best of all the fruits

Our fairing, [i.e.mango] beauty maker of the garden, fairest fruit of Hindustan.

Babur has praised the mangoes of Bengal and Gujarat for their excellent quality.[31] Plantain (sans, *Kela*, Musa sapentum) is the other fruit mentioned by Babur. He writes:

The fruit has two pleasant qualities, one that it peels easily, the other that it has neither stone nor fibre.[32]

Other fruits mentioned by Babur are *anbali* (H. *imli*, Tamarindus indica, the tamarind), *mahuwa* (Bassia latifolia), Mimusops (sans, *khrini*, mimusops kauki), *jaman* (Eugenia jambolana). About jaman, Babur writes:

its fruit is like a black grape, is sourish and not very good.

Other Indian fruits that find mention in his memoirs are *Kamrak* (Averroha carambola), jack fruit (H. *Kadhil*, Artocarpus Integrifolia), Monkey Jack (H. *Badhal*, artocarpus lacoocha), lote fruit (sans, *Ber*, Zizphyus jujube), *karunda* (carissa carandas, the corinda), *paniyala* (Flacourita cataphracta), people called it *phalsa* also. *Gular* (the clustered fig), *amla* (phyllanthus emblica, the mayrobalan tree), *chirunji* (Buchanania latifolia), date palm (P. *Khurma*, phoenix dactylifera), coconut palm (P. *nargil*, cocos nucifera), *tar* (Borassus flabeliformis the palmyra-palm), oranges (Ar. *Naranj*, citrus aurantium). But according to Babur the oranges of Lamghanat, Bajaur, and Sawad were better than those of Hindustan. Lime (B.*limu*, C.acidica), Citron (P.*turunj*, C. *Medica*), Sangtara, the fruits resembling orange, janbiri lime, another orange like fruit, galgal, *sada-fal, Amrd-fal(amrit phal)*, the lemon (H.*Karna*, C. limonum) and *amal bid*.[33] Babur is full of praise for the fruits of Kabul. He regards the fruits of Kabul as of best quality. According to him:

Fruits of hot and cold climates are to be had in the districts near the town. Amongst those of the cold climate, there are had in the town, the grape, pomegranate, apricot, apple, quince, pear, peach, plum, sinjid, almond and walnut.[34]

He further writes:

The rhubarb of Kabul district is good, its quinces and plums are very good, so too its badrang, [35] *it grows an excellent grape known as water grape.*[36]

Mirza Haider Dughlat (1543-44) has given a detailed description of the fruits which were grown in Kashmir. He writes:

As for the fruits – pears, mulberries [sweet], cherries and sour cherries are met with, but the apples are particularly good. There are other fruits in plenty, sufficient to make one break one's resolution. Among the wonders of Kashmir are the quantities of mulberry trees [cultivated for their leaves], from which silk is obtained. The people make a practice of eating its fruit… In the season the fruit is so plentiful that it is rarely bought and sold. The holder of a garden and the man that has no garden are alike; for the gardens have no walls and it is not unusual to hinder anyone from taking the fruit. He further writes *"The fruits are especially good and wholesome. But, since the temperature be inclines to cold and the snow falls in great abundance, those fruits which require much warmth, such as dates, oranges and lemons do not ripen there; these are imported from the neighbouring warm regions.*[37]

In *Ain-i-Akbari*, a celebrated work of Akbar's reign, a long list of Indian fruits is given. Among these mulberries and *gulars* were grown during spring; pineapples, oranges, sugarcane, *bers, usiars, bholsaris, gumbhis, dephals* during winter; jack fruit, *tarkuls*, figs, melons, *lah saruas, karahris, mahuwas, tendus, pilus, barautas* during summer; and mangoes, plantains, dates, *delas, gular*, pomegranates, guavas, watermelons, *paniyalas, bangas, khirni, piyars*, during the rains.[38] According to Abul Fazl:

these fruits of Hindustan are either sweet, or subacid, or sour; each kind is numerous.[39]

Several steps were taken for the encouragement of tree plantation and for the improvement in Horticulture after the coming of Mughals. Orchards were laid out by the members of the higher classes of the society, by the emperors themselves, by princes and by nobles. The state as well as society encouraged the plantation of walled as well as unwalled gardens and orchards on an equally large scale. The sources mention a circle of thousands of big gardens around almost all the cities. Towns had the same pattern. Villages had their own share of gardens both large and small. These gardens produced fruits and pot herbs for the markets.[40] With the coming of Babur horticulture received a special attention. Babur combined two essential qualities for a gardener, a natural affinity with plants and a dedicated attention to detail. When finally settled to his life in India he still found time to write to his governor in Kabul, Khwaja Kilan, with instructions that the gardens planted there should be kept well watered and properly maintained with flowers. Whatever matter in hand whether comparing different ways of raising water from a well, choosing oleanders for their colours, or arranging for the import of fruit trees, he brought to the project the sane plan and commitment that gave him his victories in battle.[40a] He encouraged the plantation of the gardens consisted of fruit bearing and flowering trees. On the banks of the Jamuna, he raised several gardens with beautiful flowers and fruit trees. Love for fruits and gardens continued in the Mughal family. During the times of Akbar, Jahangir and Shahjahan gardening was a craze. Even Aurangzeb (1658-1707) had a lot of interest in planting of trees. He had written a few letters to his sons to be very careful in planting trees.[41] These gardens were laid down in almost all parts of the empire, but it was Kashmir where the maximum number of the gardens were laid out. Gardens gave a fillip to horticulture. Experiments on grafting and plantation of new varieties of fruits were carried out in these gardens, with too well known and beneficial effects for the variety and exuberance of fruit industry in the valley.[42] The variety of plants found in the Mughal gardens was great and, by some accounts, there were so many different types that gardens could be kept continuous bloom all the year.

The "*fruitful trees and delightful flowers*" said Terry, "*seems never to fade.*"

By Peter Mundy's account of 1632, the fruit trees in the Mughal gardens included:

Apple trees (those scarse), orange trees, mulberrie trees, etts. Mango trees, caco[coconut] trees, fig trees, and plantain trees as well as cherry, apricot, pomegranate, and guava.[43]

Della Valle too has given a vivid account of the fruit trees in the garden of Jahangir. He writes:

March the 5th we visited the King's garden again, and many other gardens, where we tasted diverse fruits and beheld several flowers unknown in Europe.[44]

About the fruits of India Edward Terry writes:

The country abounding in the musk-melons (very much better, because they are better digested there by the heat of the sun, than these with us). They have many water-melons, a very choice good fruit, and some of them as big as our ordinary pompanos, and in shape like them; the substance within this fruit is spongy, but exceeding tender and well tasted, of a colour within equally mixed with red and white and within that an excellent cooling and pleasing liquor. Here are like wise the store of pomegranates, pome-citrons, here are lemons and oranges, but I never found any there so good as I have some elsewhere. Here are dates, figs, grapes, prunelloes, almonds, cocoa-nuts and here they have the most excellent plums, called mirabolans, the stone of which fruit differs very much from others in its shape, whereon hath curiously quartered several strakes equally divided, very pretty to behold, many of which choice plums are very cordial and therefore worth the prizing, are there well preserved and sent for England.

They have to these another fruit we English there call a Planten, of which many of them grow in clusters together; they are long in shape, made like into slender cucumbers and very yellow when they are ripe, and then taste like unto a Norwich pear, but much better. Another most excellent fruit they have, called a mango, growing upon trees as big as our walnut trees; and as these here, so those trees there, will be very full of that most excellent fruit, in shape and colour like our apricots but much bigger; which taken and rolled in a man's hands when they are through ripe, the substance within them becomes like that pap of a roasted apple, which then sucked out from about a large stone they have within, is delicately pleasing unto every palate that tastes it. And to conclude with the best of all other their choice fruits, the Ananas, like unto our pine-apples... In the north-west part of empire they have variety of pears and apples.[43a]

During Shahjahan's reign a vast amount was spent on the construction and maintenance of these gardens and planting of trees. Eight lakhs of rupees were spent in a year for this purpose.[45] William Finch, who visited a royal garden at Sirhind in 1611, writes that the garden was richly planted with all sorts of fruits and flowers and a large number of cypresses trees were planted there.[46] A special attention was paid on the cultivation of the fruits trees in the gardens. Manucci, an Italian traveller of 17th century India, writes:

Round the city are fine gardens filled with various kinds of fruits, chiefly peaches, which are fine and large and in great abundance ... there are may quinces (marmelos), figs, mulberries, stoneless grapes, mangoes and melons of may kinds.[47]

Mughal emperor Muhammad Shah (1719-1748) got many fruit trees to be planted in his gardens.[48] In the later period Nawab Shujaud-Daula of Awadh (1754-1775) had laid out many mango trees and other fruit trees in his gardens at Faizabad.[49] The court and aristocracy made great endeavours to grow almost every variety of fruits and flowers in their gardens. They planted fruits brought from various regions, imported seeds as well as gardeners, laid out expansive system of irrigation, and improved certain varieties of fruits by propagating grafting technique. The attempt to grow Central Asian fruits began with Babur. Even prior to coming to Hindustan, Babur had introduced new fruits in Kabul for instance he planted *alu-balu* (a kind of plum) at Kabul. He writes:

I had the cuttings of the alu-balu brought there and planted; they grew and have done well.

He also took banana plants and sugarcane from Lahore to Kabul and planted them there.

The year I defeated Pahar Khan and took Lahore and Dipalpur, I had plantains (bananas) brought and planted there; they did very well. The year before I had sugarcane planted there, it also did well; some of it was sent to Bukhara and Badkhshan.

Babur brought some of the best musk melon plant of Kabul to India and got them planted in his garden at Agra. He also imported grapes seeds from Balkh and planted them at Agra. He writes:

A Balkhi melon-grower has been set to raise melons; he now brought a few first rate small ones; on one or two bush-vine; (butatak) I had planted in the Garden-of-eight-paradises, very good grapes had grown; Shaikh Guran sent me a basket of grapes which were not bad. To have grapes and melons grown in this way in Hindustan filled my measure of content.[50]

Mughal emperor Jahangir writes in his memoirs that most of the fruits such as melons that were not grown in India, were introduced in the reign of Babur and now those grown in India were as good as those of Iran and Turan.[51] Babur is also credited with the introduction of Persian scented rose in India.[52] Akbar also took a lot pain in improving the quality of fruits in India. According to Abul Fazl:

Skilled hands from Turkestan and Persia under his Majesty's patronage, sowed melons and planted vines and traders began to introduce in security the fruits of those countries, each in its season and with attention to its quality, which occasioned an abundance here when they were not procurable in their own.[53]

He further writes:

Through the encouragement given by His Majesty, the choicest production of Turkestan, Persia and Hindustan are to be found here. Musk Melons are to be had though out the whole year. They come first in the season when the sun is in Taurus and Gemini; (April, May, June), and a later crop when he is in the Cancer and Leo (June, July, August). When the season is over, they are imported from Kashmir, and from Kabul, Badakshan, and Turkestan.[54]

Akbar not only encouraged the cultivation of indigenous trees, but he also introduced some new trees and fruits that were not grown in Hindustan for instance he is credited with the introduction of cherry *(shahalu)* and *chinar trees* from Kabul to India.[55] Like his ancestors Jahangir also made great efforts to improve Horticulture. He had so much love for plants and trees that he ordered to cut the thumbs of servant of one of the noble who destroyed the *champa* trees. He writes:

A gardener reported that Muqarrab Khan's servant had cut down several champa trees… I was so stunned to hear this… I ordered both his index fingers cut off to serve as an example to others.[56]

Like a true gardener Jahangir felt special pleasure in tasting home grown fruits;

There was a young plant in the little garden of Ishrat-afza (joy-enchanting), he remarks, this is called nau-bar (new fruit). Every day I plucked with my own hands sufficient to give a flavour to my cups. Although they sent them by runners from Kabul as well, yet to pick them oneself from one's home garden gave additional sweetness.⁵⁷

Sweet Cherry was not grown in Kashmir before Akbar's reign, but now Muhammad Quli Afshar introduced it from Kabul by means of grafting. Jahangir writes:

Before His Majesty Arsha Ashyani's reign there were absolutely no cherries. Muhammad Quli Afshar brought them from kabul and grafted them. Now there are ten or fifteen fruit bearing trees. There were also few trees of grafted apricots. the same person spread grafting through out the land, and they are now abundant.

He further writes:

*There are pears (nashpati) of the best kind, better than Kabul or Badakshan and nearly equal to those of Samarqand. The apples of Kashmir are celebrated for their goodness. The guavas (amrud) are middling. Grapes are plentiful but most of them are harsh and inferior, and the pomegranates are not worth much. water melons of the best kind can be obtained ... There are no shah-tut but there are (tut) mulberries everywhere. From the foot of every Mulberry tree a vine-creeper grow up*⁵⁸

Mulberry trees were also grafted in Kashmir gardens to bring the fine quality of mulberry.⁵⁹ In Kabul grafted apricot trees *(zard-alu paywandi)* were planted by Mirza Muhammad Hakim, Jahangir has praised the quality of apricot very much according to him:

*The apricots from other trees cannot be compared to the ones from this tree.*⁶⁰

Muqarrab Khan, the governor of Gujarat during the reign of Jahangir, had grown the best quality of mangoes in his garden by importing good quality seeds from all over the country. Praising the mangoes of Muqarreb Khan's garden at Kairana (between Delhi and Sirhind), Mu'tamed says that Muqarrab Khan "had obtained seeds of mangoes from Dakin, Gujrat and other distant parts of which he had heard any praise, and planted them here." He adds that this garden covering an areas of 140 bighas or 33.6 hectares, contained "*a large number of trees native to warm and cold climates*".[61]

Jahangir too has praised the mangoes of Muqarrab Khan's garden.

Mangoes are not found in Hindustan past the end of the month of Tir (June- July), but Muqarrab Khan had made orchards on the pargana of Kairana, his ancestral home land, in which they can be kept and preserved somehow for an extra two months after their season so that mangoes were sent to the fruit store house in Ajmer every day.[62]

Trees bearing the better class of fruits such as quality mangoes were planted in grooves, in carefully measured rows to increase production and to improve quality.[63] Not only the emperors and nobles worked for the development of horticulture but queens and princesses actively took part in it. Nurjahan, the queen of Jahangir laid out several gardens in different parts of the empire. She spent a vast amount on the construction of avenues and plantation of trees.[64] She was so much struck by the beauty of the flowers of Kashmir that on her return from Kashmir in 1620, she not only introduced new varieties in the gardens of the plains, but flowers became an important design motif in art as well.[65] Nobles like Asaf Khan also grew different varieties of flowers and trees such as cypresses, chinars, iris, narcissus, daffodils, and roses in their gardens.[66] During the reign of Shahjahan, the honey-dew variety of melons, a native of colder regions, was successfully cultivated in the plains of North India. Muhammad Rida of Khurasan who raised the first crop was rewarded and honoured by Shahjahan.[67] Love of Shahjahan for the fruits is obvious from the incident that there was a mango tree named *Badshah Pasand* (favourite of the Emperor) at Burhanpur, and Shahjahan was very much fond of its fruit. When Aurangzeb was in Deccan, Shahjahan had asked him to regularly send mangoes to him. Aurangzeb appointed men to look after the trees and promised to obey the royal order. When the season came, mangoes were sent to emperor, but he was not satisfied either with their quality or quantity, and suspected that Aurangzeb appropriated good mangoes for his own table. Aurangzeb protested that mango crop was very bad and *Badshah Pasand* yielded very few mangos. But the emperor's suspicion as not allayed and he wrote a letter that since he had not received mangoes in sufficient quantities, he would appoint his own men next year and they would send mangoes directly to him.[68]

Earlier the practice of grafting was restricted to imperial gardens only for the reason of prestige, but Shahjahan lifted this ban for both "the select and the masses". Remarkable results are said to have followed from its wider application. The quality of the Oranges, the *Sangtara*, *kola* and *narangi* was greatly improved. Quality of mangoes also improved by the system of grafting. The Portuguese in India were probably the first to create varieties of mangoes by grafting. Grafted mangoes are reported from Bengal only in the 18th century, and the practice might have been brought there by Portuguese.[69] "*Alfonso*", the famous grafted mango of western coast in Goa is ascribed to one Nicole Afonco of Goa. Manucci writes:

I may mention that the best mangoes grow in the island of Goa. They have special names, which are as follows mangoes of Niculao Affonso, Malaiasses (? Of Malacca), carreira branca (white carreira), carreira vermehla (red carreira) of Conde, of Joani parreira, Babia (large and round), of Araup, of porta, of secreta, of Mainato, of our lady, of Aguade lupe. These are again divided into varieties with special colour, scent and flavour. I have eaten many that have the taste of peaches, plums, pears, and apples of Europe.[70]

Francois Bernier, a French traveler of 17th century India, has written that pateques or water-melons were cultivated by importing good quality seeds and were given much care and expanse.[71] Masudat Munshi in his *Adab-i-Alamgiri* refers that the flowers '*lala*' and '*yashmine*' and fruits '*nashpati*' and '*shakhtalu*' were grown in the spacious and beautiful gardens of Lahore.[72] *Sujan Rai Bhandari* in his

Khulasat ut Tawarikh has described the melons and vines of Lahore as good as those of Persia and Turkestan.[73] Praising the mangoes of Bengal Gulam Hussain Zaidpuri writes

mango is the best fruit of the land. In some places a type of mango is produced which is big, sweet and without fiber.[74]

There are references to the construction of gardens during the reign of Muhammad Shah in which different varieties of trees and flowers especially lotuses in the pools were grown.[75] During Safdar Jang's reign (1748-1752) there is reference of a incident which shows that cutting of trees was regarded as a serious crime. A camel driver in the service of Mughal captain (i.e. a Persian – Turk Soldier of Safdar Jang) cut down a tree growing before the gate of Inyat Khan, an officer of the *wazir*, who chastised him severely for it.[76] In the later period many provincial rulers worked for the development of horticulture. The Nawabs of Oudh and their officers planted many mango gardens in the city of Lucknow and is environs. Even today also Lucknow is famous for its *Sufaida* – a light yellow sucking variety of mango which is unrivalled in taste.[77] Numerous grooves were planted in Lucknow in which excellent mangoes, guavas, oranges, pomegranates, jack and *bers* (zizy phusjujuba) were grown.[78] Even the French and English who settled in India during the reign of Mughals also worked to improve the quality of fruits. French laid down vineyards in Pondicherry and were rewarded by a surfeit of delicious grapes.[79] Later in 18th century in a different way, gentry and land holders also intervened in the production of the valuable fruit and vegetable crops. It was the Muslim service of gentry of the small towns and large villages of Awadh who planted and supervised the growth of mangoes, guavas and other fruit trees planted around their home market towns. Specialist market gardeners and fruiter castes (*malis, kacchis* and *koeris*) were given leases of grooves but the gentry retained an interest in improvement and marketing.[80] In 1797, near Kanpur, an island of relative agricultural prosperity was created in the midst of the depressed middle Doab, and the intensive fruit and tobacco cultivation which had been fostered by the Nawab of Farrukhabad received impetus from the growth of the British stations at Fatehgarh.[81]

Several new fruits were also introduced in the realm so that excellent melons, watermelons, peaches, almonds, pomegranates began to be commonly grown in the empire.[82] Various new species of fruits were introduced from the New World through the agency of Portuguese. The most valuable was the pineapple *(annnas sativa)*, which spread through the length and breadth of India with striking rapidity. Grown in the beginning in the Portuguese possessions on the western coast (Goa), it had by the end of the 16th century became common enough in Bengal, Gujarat and Baglana and was noticed among the important productions of these regions. In 1660 pineapple of a very good quality was found growing in Assam.[83] Pineapple figures prominently among the Indian fruits described by Abul Fazl. He calls it *Kathal-i-safari* or the traveling jack fruit and refers to its cultivation in Baglana, a mountainous tract between Surant and Nandurba.[84] During the reign of Jahangir many thousands of pineapples were produced every year in the imperial garden of Agra. Jahangir writes in his memoirs

One of the fruit is the one called ananas [pineapple]. It is to be found in the Franks' ports[85] *It is extremely good smelling and tasting. Several thousands are produced every year in the Gulafshan garden in Agra.*[86]

Muhammed Saleh Kamboh in his work *Amal-i-Saleh* has written that *ananas* was abundantly grown at Surat and Baglana.[87] *Ananas* was also reported to be grown in certain parts of Maharashtra in the reign of Shivaji.[88] Bernier has also referred to the cultivation of pineapple in Bengal.[89] Guavas and custard apples which grow so profusely today in certain parts of the sub-continent were introduced by the Portuguese and their cultivation was encouraged by the Mughals.[90] Papaya and Cashew-nut were introduced from the same source but took more time to spread. Linschoten had already found the cashew-nut growing in the Portuguese possession and noted that it had been transplanted from Brazil. Thevenot noticed it growing along the route from Surat to Aurangabad.[91] Della Valle tasted papaya and cashew-nut fruit in Daman in 1623. He writes:

In Daman I first tasted at Father Rector's table many strange Indian fruits… brought into East India from Brasil or New Spain namely papaya (papaya), cagiu (Kaju or cashew), Giambo (jambo fruit), mango or amba and ananas all which seem'd to me passably good; and though of different taste, not inferior to ours of Europe, especially papaya, which is little esteem'd in India, … in shape and taste it much resembles our melons, but is sweeter, and consequently to me seem'd better.[92]

A new variety of grafted mangoes called *"Alfonso"* was also introduced by Portuguese. *Manucci* has referred about their cultivation at Goa.

They gave them special names, taken from the first person to have good mangoes of that kind. Then they speak of the mangoes of Niculao Alfonso, which are the largest and the best; Malaisses mangoes, and careiras mangoes.[93]

Fruits like *lichi, chiku and lokat* were introduced in much later period almost in 19th century by the Europeans.[94]

As regard the revenue accruing to the state from this source, fruit gardens were treated in a special manner. A flat rate of rupees 2 – ¾ per *bighas* was charged, even if the trees did not bear fruit. Exception was made in case the orchards were planted with grapes and almonds, in which case the demand was realised only when the plants bore fruit. Sometimes if it was found that the orchard was not sufficiently productive, a 5th or a 6th of the net produce was claimed. If the maintenance of the orchard lost more than the value of the yield no change was made.[95] Akbar remitted cess on the fruit trees, *Sar-i-darakhti.*[96] Jahangir whose taste for choice fruits was notorious, states that fruit trees were and had always been, free of any demand for revenue, and that a garden planted on cultivated land was forthwith exempted from assessment.[97] Jahangir writes in his memoirs

God be praised that in this age-enduring state no tax has ever been levied on the fruit of trees, and not levied now. In the whole of the dominion not a dam nor a grain (habba) on this account enters the public treasury, or is collected by the state. Moreover, there is an order that whoever makes a garden on arable land, its produce is exempted.[98]

But from the number of documents belonging to Aurangzeb's reign it seems clear that tax was then being levied on all orchards, except for those containing graves or yielding no profit. The quantity of fruits was assessed per tree: a 5th of it was taken from the Hindus and 6th from the Muslims.[99]

Among various *Karkhanas*, a special department for fruits existed which was called the *Mewah Khana*, for the supervision of the supply, grade and the staff of the fruitry.[100] During Akbar's reign, for the *Mewah Khana* delicious, tender and sweet smelling musk melons, different varieties of grapes and many other fruits such as pomegranates, apples, peaches, guavas were supplied from Kashmir, Kabul, Badakshan, Zablistan and Samarqand.[101]

In the *Mewah Khana* the fruits were marked according to their degree of excellence, melons of first quality were marked with a line drawn round the top; those of second; with two lines; and so on. Whenever Akbar took wine or opium, the servants used to place before him stands of fruits; usually he ate little, most of them was bestowed as *"Alush"* to others. In this department *mansabdars*, *Ahadis* and other soldiers were employed; and the pay of a *piyadah* (foot soldier) varied from 140d. to 100 d. per month.[102] In Jahangir reign the same supply flowed in the *Mewah Khana*. Some new fruits were added to the list, such as *sahabi* and *habshi* grapes,[103] and melons from Yazd and Kariz.[104] The latter used to arrive in great numbers as much as 1500 in one batch. They were usually distributed by emperor among the ladies of Harem and the servants of the state.[105] Although the distance was 1400 *kos* and one *qafilah* (caravan) took about 5 months to come, yet the fruit arrived "very ripe and fresh".[106] *Bangash apples* were sent by the *Dak Chawki* to the imperial tables by the officers. They were of excellent flavour and arrived very fresh. Jahangir liked them immensely and was greatly pleased in eating them[107] Oranges were sent from Bengal by the same method. It was the duty of the department to see that they were not spoilt in transaction. In spite of being *"Very delicate and pleasant"*, they were brought *"by post as much as necessary for the private consumption"* of Emperor, passing hand to hand.[108] In Ahmedabad a regular department existed to supervise the fruit tree cultivation in the gardens of the city. And under its supervision the suburban area of Ahmedabad covered by the gardens, appeared to have been much larger than that of imperial capitals Agra, Delhi and Lahore.[109] Ali Mohammad *Khan* in *Mirat-i-Ahmadi* writes that the fruits and vegetables had certain criterion of weight, determined for sale. The sellers brought these items in a heaps or bundles of fixed weight to markets. He mentions that most of the items were sold at a price of 42 *Asar* per month; others have more or less. The fruits were brought by lots of hundreds, were one hundred per *rupee*, except the banana where actual of 120 was counted as 100. May be; it was to compensate for the loss during transit or due to spoil in storage.[110] In the later period there are references to the existence of this department, during the reign of Muhammad Shah there is a reference that Muhammad Ishaq Khan was promoted, from he superintendent of the royal gardens at Delhi to the inspector of Crown Prince's contingent. [111]

Not only fruit cultivation but plantation of different varieties of flowers was also encouraged. In all the Mughal gardens there were beds of the flowers known for their beauty and fragrance. The gardens laid out by the aristocracy adjoining their mansions used to be so planned as to consist of both fruit bearing trees and sweet scented attractive blossoms. The immigrant Muslims had introduced some flowers of the Islamic countries for instance roses.[112] Babur has praised the flowers of India, he writes that *"there is a great variety of flowers"* in Hindustan. Among the flowers he has mentioned are *jasun* (Hibiscus rosa sinensis), *Kanir* (oleander), *kiura (the screw pine),* he writes that that *kiura* has very agreeable perfume. Other flowers mentioned by him are Jasmine; the white jasmine is called *champa*

it is larger and more strongly scented than our yasman flower.[113]

Babur also took some red oleanders from Rahim-dad's flower garden at Gwalior and got them planted at his garden in Agra.[114] In *Ain-i-Akbari*, Abul Fazl has given a long list of the flowers grown in India. He writes:

it is really too difficult for me ignorant as I am, to give a description of the flowers of this country, he further writes: *Gardens and flower beds are every where to be found*[115]

He has divided the flowers grown in India in different categories. Those were 1. fine smelling flowers, 2. flowers notable for their beauty, 3. Various Irani and Turani flowers which were introduced and grown in India.[116] Fine smelling flowers include the *sweti, Bholsari, chambeli, ray-bel, mongra, champa, ketki, pedal, juhi, niwari, nargis, kewara, chalta, gulal, tasbih, gulal, singarhar,* violet, *karma, kapurbel and gul-i-zafran*. He has also mentioned the seasons in which these flowers were grown.[117] Second category of flowers includes: *Gul-i-Aftab, Gul-i-Kawal, jafari, gudhal, rattan-manjani, kesu, senbel, ratanmala, sonzard, Gul-i-malti, karnphul,* the *karil, kaner, kadam, nagkesar, surpan, siri khandi, jait, champala, lahi, Gul-i-karaunda, dhantar, Gul-i-hinna, dupahriya, bhunchampa, sudarshan, kanglai, sirs, and san.*[118] Various Irani and Turani flowers which were introduced in India were *Gul-i-Surkh,* the *nargis,* the violet, *Yasman-i-kabud,* the *susan*(the *iris),* the *rayhan* (Sweet basil), the *rana,* the *zeba,* the *shaqayiq,* the *Taj-i-khurus,* the *qalgha,* the *nafarman,* the *khatmi,* etc.[119] Abul Fazl is full of praise for the flowers of Kashmir.

The flowers are enchanting and fill the heart with delight. Violets, the red rose and wild narcissus cover the plains. He further writes: *Tulips are grown on the roofs which present a lovely sight in spring time.*[120]

Praising the flowers of India Jahangir writes:

From the excellencies of its sweet scented flowers one may prefer the fragrances of India to those of the flowers of the whole world. It has many such that nothing in the whole world can be compared to them.[121]

Of the flowers described by Jahangir are *Champa* (Michelia Champaca); he has regarded this flower of exceedingly sweet fragrance; *kewra* flower *(pandanus odoratissimus), raebel, mulsari* (Mimusops Elengi), About *mulsari* he writes that the tree is very graceful, symmetrical and shady and the scent of the flowers is very pleasant, others mentioned by him are *ketki, chambeli* (jasminum grandi florum).[122] Not only the rulers but the nobles also made great endeavours to encourage the cultivation of flowers. Asaf Khan grew every variety of flowers in their gardens.[123] Jahangir was very much struck by the flowers of Kashmir.

Kashmir is a perennial garden... Red roses, violets and narcissi grow wild; there are fields after fields of all kinds of flowers...

On his visit to Kashmir in 1620, he observed:

Sweet smelling plants of narcissus, violet and strange flowers that grow in the country came to view. Among these flowers I saw one extraordinary one. It had five or six orange flowers blooming with their heads downwards. From the middle of the flowers there came out some green leaves, as in case of the pine-apple[124]

Flowers were grown with so much care that even in the end of season flowers like jasmine were available in the Nurmanzil garden of Kashmir.[125] *William Finch* who visited Asaf Khan's garden at Lahore writes that different varieties of flowers such as roses, stockyellow-flowers, marigolds, wall-flowers, ires, pinks, white and red and several sorts of other Indian flowers were grown there.[126] Della Valle has mentioned that he saw various flowers in the emperor

Jahangir's garden which were unknown in Europe, he had made a special mention of *champa* flower.[127] The cultivation of roses seems to have received strong impetus in the 17th century because of the increased demand of rose water and the newly invented rose essence (*itr-i-Jahangir*).[128] Efforts were made to grow roses in those areas where they did not grow much well especially at Ahmedabad. As Jahangir has written:

On Wednesday I went to the Fath garden to see the roses. One whole bed was in full bloom. Roses do not grow much in this land, and to see so many in one place was an opportunity not to be missed.[129]

Plantation of flowers also received an impetus during the reign of Shahjahan. The flowers which were planted on the sides of the Dal lake in Kashmir were so beautiful that *"they were enough to be wilder the gaze."* Different varieties were planted there but the most notable was *scarlet lotus*, which was called *kanwal* in Hindustan. In Kashmir were planted the bulbs of *Lala Chughasu*, one of the finest species of tulip which bear gorgeous flowers.[130] About the flowers of Kashmir Perter Mundy writes:

Roses... French mariegolds aboundance, poppeas redd, carnation and white; and diverse other sorts of faire flowers which wee knowe not in our parts, many groweinge on prettie trees, all watered by hand in tyme of drought, which is 9 monethes in the yeare.[131]

In Orissa *kewra flower* and *nasrin*, a very delicate flower with nice smell, were grown.[132] Manucci writes that nearly a crore of rupee were spent by the Mughals for the cultivation of flowers and extraction of scents.[133] Sujan Rai has given a detailed description of the flowers of Kashmir *the roses, the tulip, and especially the narcissus, grow wild in abundance.*[134]

During the reign of Aurangzeb, a wide variety of flowers were grown which were diverse in colours and in their aroma, like rose, *jasmine, narcissus, sosan, lala, zanbag, banafsha, rihan, rana, ziba, nafarman, taj, khrus, qulfa, abbasi, khatmi, arghuan, sadbarg* and *dandi* etc. All these flowers were found in Iran and Turan also. But there were certain flowers which were found in India only and there number was beyond count or description. *Sevati, Champa, mogra, motia, ramansej, kanwal, aftabi* and many others. All these flowers were of very attractive and diverse colours with fine and refreshing aroma. However *kewra* and *ketki* were exceptions in delicacy of their smell and refreshness. So much so that just a handful of these provided beautiful aroma for a gathering of 100 of persons. They filled not only the house but the whole locality with aroma. These flowers were not found in Iran and Turan. The delicacy of *raibel, chambeli*, aroma and their refreshing qualities were well known all over the world. These flowers were used on all the happy occasions like wedding, birth, and religious ceremonies.[135] Bernier has regarded that most of the European flowers were grown in Kashmir.[136] Flower cultivation was encouraged in various regions like Aurangabad, Malwa, Surat, Khandesh etc.[137] At Delhi various sweet smelling flowers as roses, water lilly, Egyptian willow, narcissus, jasmine were found.[138] During the reign of Bahadur Shah Jaffar (1707-1712), the festival of *Phool Walon ki Sair* (festival of flower sellers) was celebrated. The King would himself go to the *"Jharna"* (a beautiful garden with cascades and fountains, originally built by Feroz Tughluq) to celebrate the festival.[139] In the late 18th century during the British rule there are references to the cultivation of the flowers like rose, *chambeli*, jasmine.[140] At Madras British grew all sorts of flowers such as jasmine for their beauty and delight in their gardens.[141]

Apart from planting trees in the gardens, a large number of the fruit trees and shady trees were planted on the both sides of the road either directly by the state or else by some wealthy philanthropists. Such shady trees were known as *khayaban* or avenues.[142] These were planted to provide fruit and shelter to the tired way fares. Practice of plantation of trees on the roadside was encouraged by Sher Shah Suri. On the either sides of his four great roads. i.e. from Sonargoan in Bengal through Agra, Delhi and Lahore to the Indus, from Agra to Mandu and one from Agra to Jodhpur and Chitor and one from

Lahore to Multan, were planted fruit trees.¹⁴³ Akbar continued this policy and issued orders for the plantation of trees on both sides of the roads. He caused the whole road from Agra to Lahore to be planted with an alley of trees, beautiful for its length and verdure.¹⁴⁴ Father Monserrate who visited India in the reign of Akbar writes that the roads were planted down the middle with beautiful green trees casting grateful shades.¹⁴⁵ Jahangir had too much love for trees, that in *Shahrara* garden at Kabul he gave the names *Farah-bakhsh* (joy giver) and *Saya-bakhsh* (shade giver) to his two favourite trees.¹⁴⁶ He also issued *farmans* to zamindars and other officials to construct *serias* and plant shade trees on road sides.¹⁴⁷ He got the trees to be planted on the road from Agra to Bengal. He writes in his memoirs:

*Previously to this, according to order, they had planted trees on both sides from Agra as far as the River of Attock (the Indus), and had made an avenue, and in the same way from Agra to Bengal.*¹⁴⁸

William Finch, who travelled through this road writes,

*All along on both sides the way from Kabul to Agra, a reasonable distance, the King caused trees to be planted to shade the way in remembrance of this exploit, and called this place Fatehpur, that is Heart Content which for his birth was named so by his father Akbar.*¹⁴⁹

About the plantation of trees on this road De Laet writes:

*the road is bordered by trees on either side which bear fruit something like mulberry.*¹⁵⁰

Peter Mundy informs us that Jahangir had ordered a road to be built from Patna to Agra for the comfort of the travellers, and found it covered with shady trees on either side *"each at a distance of 8 to 10 steps and the rank from side to side about 40."*¹⁵¹ Thomas Coryat, who visited India from 1612 to 1617 writes that the road from Lahore to Agra was good and planted along its sides with rows of trees.¹⁵² The facilities on the roads and shade of trees used to lessen the discomfort of long journeys, some of which lasted for months.[153] Manucci has given the account of the plantation of the trees on the road from Multan to Allahabad by Jahangir. He writes;

The King was very fond of carrying out works for the benefit of the public and the adornment of his kingdom. To this intent he issued orders to have trees planted on the royal highways from one end of the country of the other, commencing at the city of Moltan (Multan), and going as far as the city of Ilavas (Allahabad) – that is, for five hundred and thirteen leagues.[154]

Tavernier has also made a mention of the tree plantation on the road side. He remarked:

that nearly all the way from Lahore to Delhi, and from Delhi to Agra, is like a continuous avenue planted through out with beautiful trees on both sides, which is very pleasant to view…[155]

Tom Coryat enjoyed his "Long Walk" of 400 miles in length *"shaded by trees on both sides".*[156] Delaet has given the account of the tree plantation from Bara Palah bridge to Humayun's tomb at Delhi. According to him:

Not far from these ruins a branch of the river Jemini or Gemena (yamuna) is crossed by a stone bridge [157] *from which a broad avenue, thickly shaded by tall trees, leads to the tomb of Humayun…* [158]

William Finch has also mentioned about it, he writes:

A little short is a stone bridge of eleven arches, over a branch of Gemini[the Jumna]; from hence a broad way shaded with great trees leading to the sepulcher of Hamaron [Humayun]…[159]

Trees were not only planted on the roads and gardens but the nobles used to plant them in their courtyards also. De laet writes:

In the courtyard they (nobles) have a tank and trees to mitigate the heat[160]

Bernier who visited India during the reign of Aurangzeb writes that In Agra in the houses of *Omrahs*, rajas and others were planted big green trees [*buta*], everyone encouraged the plantation of trees in his garden as well as courtyard for shade and beauty.[161] In Kashmir a large number of trees were planted especially around Dal lake. The lake was surrounded by the large leafed trees, planted at the interval of two feet. The largest of these trees might be clasped in man's arms, but they were as high as the mast of a ship[162] Mughal Emperor Muhammad Shah also planted trees at his garden at Delhi.[163] In the later period the Nawabs of Awadh, and Kings of Jaunpur and of the Deccan did the same.

They beautified the neighbourhood of their own favourite residences, made roads to the country seats… and planted lines of trees.[164]

With the efforts of the Kings and Aristocracy there was an increase in the production of fruits through out the empire which resulted in an increase in their availability. Fruits green and dry, of different varieties constituted an important item of the diet consumed especially by the well to do classes. Melons and grapes had become very plentiful and excellent; and watermelons, peaches, almonds, pistachios, pomegranates, etc. were found every where. And with the conquest of Kabul, Qandhar and Kashmir, loads of fruits were imported; through out the whole year the stores of the dealers always remained full, & there was increased supply in the bazaars.[165] Father Monserrate has given a long list of the fruits grown in India viz.,

The grape, the peach, the mulberry, the fig (in few places), and the pine tree. The whole country bears pomegranates in abundance. The cotonian apple, the pear and similar fruits are imported from Persia.[166]

Almost all varieties of fruits were grown in India. Kashmir and Kabul produced fruits of colder climates,[167] while the rest of the empire was better endowed with fruits of the warmer regions such as bananas, oranges, mangoes and guavas, etc[168] About the fruits of Kashmir Mohammad Amin Qazvini writes:

From out of the fruits of the colder regions there is hardly any fruit that does not grow in Kashmir... Shahalu of this place has no rival either in beauty or color or in the delicacy of taste... Apples are very large and very tasty. Nashpati grows in large quantities. It is very juicy and delicate. Pomegranate are average. Musk Melon, if it survives the natural calamities, is as good as the one from Khurasan. Water Melon... is large and very reddish inside. It is very sweet. Kashmir has large varieties of grapes. However due to humid air, even though large in size, it is not sweet. Unnab, Badam and Gargan grow very well.[169]

A large varieties of fruits very fine in quality were grown in different parts of empire. Abul Fazl writes:

Musk melons come in season, in Hindustan, in the month of Farwardin [February-March] and are plenty in Urdibihish [March-April]. They are delicious, tender, opening, sweet-smelling, especially the kinds called nashpati, babashaykhi, alisheri, alcha, barg-i-nay, dud-i-Chiragh, etc They continue in season for two months longer[170]

About the musk-melons of Kashmir Mohammad Saleh Kamboh Writes:

Musk melons of Kashmir if it survives the two calamities i.e. Hailstones and worm-eating, and comes out as a whole, is extremely tasty.[171]

About the grapes Abul Fazl Writes

Various kinds of grapes are to be had from Khurdad (May) to Amurdad (july) whilst the markets are stocked with Kashmir grapes during shariwar (August-sept.). Eight seers of grapes sell in Kashmir for one dam, and the cost of transport is 2 rupees per man... From Mihr (Sept.) till urdibihist grapes comes from Kabul together with cherries, which His Majesty calls Shahalu, seedless pomegranates, apples, pears, quinces, guava, peaches, apricots, girdalus, and aluchas, etc. many of which fruits grow also in Hindustan. From Samarqand even they bring melons, pears and apples.[172]

Father Monserrate is full of praise for the fruits grown at Nagarkot (Kangra). *This district produces in abundance those fruits and crops, which are characteristic of Spain and Italy, but which are not found else where in India*[173]

Abul Fazl has divided the fruits of India into sweet fruits *(Mewa-i-Shirin Hindi)*, dried fruits *(Mewa-i-Khuskhi)*, sour fruits *(Mewa-i-tars)* and fruits some what with acid *(Mighush)*. Each kind was numerous and some of them tasted well. He has given the price of various Turani and Indian fruits. [174]

TURANI FRUITS

FRUITS	*PRICES*
Arang Melons, Ist quality, at	2 ½ R
Do., 2nd and 3rd do., at	1 to 2 ½ R
Kabul Melons, Ist do., at	1 to 1 ½ R
Do., 2nd do., at	¾ to 1 R.
Do., 3rd do., at	½ to ¾ R.
Samarqand apples 7 to 15 for	1 R.
Quinces, 10 to 30 for	1 R.
Pomegranates per man,	6 ½ to 15 R.
Guavas, 10 to 100 for	1 R.
Kabul and European apples 5 to 10 for	1 R
Kashmir grapes, per man	108 dams
Dates, per ser	10 d.
Raisins (kishmish), do	9 d.
Abjosh (large raising), do	9 d.
Plums do	8d.
Khubani(dried apricot), per ser	8 d.

Qandhar dry grapes, do	7 d.
Figs, per ser	7 d.
Munaqqa, do	6 ¾ d.
Jujubes, do	3 ½ d.
Almonds without the shell, do	28 d.
Pistachios with shell, do	9 d.
Chilghuza nuts, per ser	8 d.

THE SWEET FRUITS OF HINDUSTAN

Mangoes, per 100, up to	40 d.
Pine-apple, 1 for	4 d.
Oranges, 2 for	1 d.
Sugarcanes, 2 for	1 d.
Jack Fruits, 2 for	1 d.
Plantains, do	1 d.
Ber, per ser	2 d.
Pomegranates, per man,	80 to 100 d.
Guavas, 2 for	1 d.
Figs, per ser	1 d.
Mulberry, do	2 d.
Custard-apple, one for	1 d.
Melons, per man	40 d.
Water-melons, one	2 to 10 d.
Khirni, per ser	4 d.
Mahuwa, do	1 d.
Dates, per ser	4 d.
Bholsari, per ser	4 d.

DRIED FRUITS

Coconuts one for	4 d.
Dry-Dates, per ser	6 d.
Walnuts, do	8 d.
Chiraunchis, do	4 d.
Makhana, do	4 d.
Supyari, do	8 d.
Kaulgatta, do	2 d.

Dates, walnuts, *Chiraunchis*, and *Kaulgatta* are in season during summer, and coconuts, *makhanas*, and *supyaris*, during winter.

SOUR FRUITS

Limes, 4 up to	1 d.
Amalbet, do	1 d.
Galgal, 2 up to	1 d.
Awla, per ser	2 d.

Limes and *awals* are to be had in summer, the others during the rains.

FRUITS SOMEWHAT ACID

Ambli, per ser	2 d.
Badhal, 1 for	1 d
Kamrak, 4 up to	1 d.
Narangi, 2 up to	1 d
Mountain grapes	
Jaman, per ser	1 d
Phalsa do	1 ½ d
Karaunda do	1 d
Karna, one for	1 d.

Kamraks and *narangis,* are in season during winter; *ambles, badhals,* mountain grape, *phalsas, labhiras,* during summer, and *kaits, pakars, karmar, jamans, karaundas, jhanbiris,* during the rains.[175] All these fruits were available in large quantities in the markets during the whole year.

Jahangir too notifies an increase in the production of fruits. He writes:

Melons, mangoes, and other fruits grow well in Agra and its neighbourhood ... In the reign of my father (Arsh-ashyani) many fruits of other countries, which till then were not to be had in India, were obtained there. Several sorts of grapes, such as the sahibi and the habshi and the kishmishi, became common in several towns; for instance in the bazaars of Lahore every kind and variety that may be desired can be had in the grape season.[176]

Pelsaert who visited India during Jahangir's reign has mentioned about the scarcity of fruit trees in India and he writes that all the fruit was imported from Kandhar or Kabul –

No apples, pears, quince, pomegranates, melons, almonds, dates, raisins, fieberts, pistachios, and many other kind. Great and wealthy amateurs have planted in their gardens Persian vines which bear seedless grapes, but the fruit does not ripen properly in one year out of three. Oranges are plentiful in December, January, and February, and are obtainable also in June and July. They are very large, especially in the neighbourhood of Bayana. Lemons can be had in large quantities. The other fruit have too little taste, and are though too little of, to be worth mentioning.[177]

About the fruits of Kashmir, he writes that in Kashmir were produced many kinds of fruits as apple, pears, walnuts, etc. but he has regarded them as inferior in flavour to those of Persia and Kabul.[178] As regard the Pelsaert views about scarcity of fruits in India his observations is untrue because contemporary Persian sources as well as the Travel accounts are full of the description about the fruits of India and corroborate the fact that a large number of fruits were produced in India. Abul Fazl has given a long list of Indian Sweet fruits *(Mewa-i-Shirin Hindi)*, dried fruits *(Mewa-i-Khushki)*, sour fruits *(Mewa-i-tars)* and fruits somewhat acid *(Mighush)*.[179] De laet has mentioned about the cultivation of fruits like *Kanwal kakri* and *singhara* (water chestnut) at Agra. He regards both these fruits as productive of cold.[180] William Finch too mentions about the cultivation of *Singharas* at Agra. He writes that its fruits were

green, soft and tender, white (inside), metish in taste and exceedingly cold in its effect.[181]

About the fruits of India Edward Terry (1616-19) writes that the country was abounding with muskmelons. One could also find water-melons, pomegranates, lemons, oranges, dates, figs, grapes, cocoanuts, plantains, mangoes, pineapples, pears, apples, etc. He has regarded *ananas* as the best fruit.[182] William Finch (1608-11) has also mentioned about the availability of a wide variety of quality fruits in India like apples, *toot* (tut, mulberry) white and red, almonds, peaches, figs, grapes, quinces, oranges, lemons, pomegranates, etc.[183] Della valle has praised the grapes of Surat, he has regarded them as ripe, sweet and good even in the month of February like *Uva-jugliatica* or the early july grapes of Italy.[184] Among the fruits grown at Goa mentioned by Manucci are mangoes, jack fruit, he has mentioned three types of jack fruit – *barca, papa, and pacheri* jack fruit. These fruits were very large and some weighted about 80 pounds. Other fruits mentioned by Manucci are pine-apple, he has mentioned that good quality pineapples were grown in Bengal; other are myrobalans especially grown in Gujarat and were exported from India to Europe.[185] Among the fruits of Kabul Jahangir has regarded the grapes as very good especially the varieties *sahibi* and *kishmishi*. He has also regarded the *shahulu (cherries)*, apricots and peaches of Kabul of excellent quality.[186] He mentions that peaches were very big, as big as a quince and a single peach weighed about 63 *Akbari* rupees or 60 *tolas*. About peaches he writes:

Of all varieties from trees that were eaten in Kabul none was better than this.[187]

Manucci has also praised about the peaches especially of Lahore. According to him the peaches were very fine and large and each one weighed about 13 ounces. Among the other fruits available at Lahore were quinces, figs, mulberries, stoneless grapes, mangoes and melons of many kinds.[188] Jahangir has regarded mango as the best fruit of India. About mangoes he, writes *"My particular favourite is the mango."*[189] He has regarded the mangoes of Burhanpur of very good quality and a single mango weighed about fifty-two and a half *tolas*.[190] Jahangir has praised the cherries of Kashmir as well as of Kabul. About cherries of Kabul he writes:

The cherry, to my taste, is the most delicious of all the fruits, Four trees have born fruit in Nurafza garden. I named one of them shirinbar [of sweet fruit], the second khoshguvas [of good taste], the third, which produced the most fruit of all, Purbar [full of fruits], and the fourth, which had the least fruit, kambar [of little fruit]. Every day I picked with my own hands just enough to have as a relish with my cups. Although cherries were being brought from Kabul by post, one derives a special pleasure from picking them really fresh with one's own hand from the garden. The cherries of Kashmir are not inferior to those of Kabul, in fact, they are even larger. The biggest one weighed a tank and five surkhs.[191]

The reference of mango cultivation in Patna is obvious from Peter Mundy's account, who visited Patna in 1632. According to him the road from Naubatpur [about 13 miles s.w.] to Patna was full of *"a million of mango trees in plots and grooves"* on both sides. There was *"noe wast ground all the way but full of mangoes trees..."*[192] During the reign of Shahjahan all varieties of fruits and flowers were grown in Kashmir. Some indeed were in comparatively superior, whilst a small number, owing to the humidity of the atmosphere, had less flavour and were only of middling quality. The apples, quinces, pears and peaches thrived well and they bore first rate fruit so much that an apple was found equal in weight to 60 *miskals*, a pear to 50, and a peach to 60, on their being weighed in royal presence.[193] The musk melons which were grown in Kashmir if they did not get devoured by insects, they strongly resembled the choice species which were grown in Khurasan. The water melons were unsurpassed and walnuts were more abundant in Kashmir than any where else in the world. The cherries, almonds, grapes, and all sorts of plums were most excellent.[194] Not only Kashmir but other parts of country for instance Lahore also produced abundant fruits and flowers.[195]

Bernier was quite impressed with the mode of display of fruits in the fruit market of Delhi. During the summers

the shops were well supplied with dry fruits from Persia, Balkh, Bokhara, and Samarqand, such as almonds, pistachios, walnuts, raisins, prunes, and apricots

and in the winter with

excellent fresh grapes, black and white, brought from the same countries, wrapped in cotton; pears and apples of 3 or 4 sorts and those admirable melons which last the whole winter.[196]

The imported fruits despite their plentiful supply, were quite expansive, for instance, the price of a single melon could be as high as two and a half crown. Only the *umrahs*, could afford 20 crowns daily on these for breakfast alone[197] The local mango supply was supplemented by the imports from Goa, Golconda and Bengal on the account of their superior quality.[198] Emperor Aurangzeb used to get his mangoes regularly from Allahabad, Malwa and Khandesh, pomegranates from Jodhpur and Thatta and other fruits from Gujarat.[199] Bernier has regarded the melons grown in India as of inferior quality and according to him:

There are no means of procuring good ones but by sending to Persia for seed, and sowing it in ground prepared with extra ordinary care, in manner practiced by the grandees. Good melons, however are scarce, the soil being so little congenial that the seed degenerates after the 1st year.[200]

Bernier has praised the *pateques* or water-melons, these were available all the year round. But those grown in Delhi were not good as they were soft, without colour and sweetness. Good fruit was found only among the wealthy people as they imported seed and cultivated it with much care and expanse[201] Bernier's account is full of praise for the fruits and flowers of Kashmir. According to him the whole ground of Kashmir was enamelled with European flowers and plants and covered with apples, pears, plums, apricot and walnut trees, all bearing fruit in great abundance. Even the private gardens were also full of fruits like melons, red beets, radishes, etc.[202] First rate melons, apples, peaches, apricots, plums and water-melons were grown in Kashmir and also grapes were grown in large numbers.[203] In Bengal various types of fruits like citrons, lemons, oranges, *ananas*, coconut, small myrobalans, etc. were grown.[204] Bananas grew in large quantities in Bengal. Although grapes and melons did not grew there but people sowed seeds and they grew also, but fruits were not good. Sugarcane grown in Bengal was very good, delicate, extremely sweet but black, red and white in colours. Mango was the best fruit of Bengal and in some places; a type was produced which was big, sweet and without fiber.[205] Regarding the quality of fruits grown in India, Bernier has regarded them as inferior as compared to those of Europe and according to him this was mainly due to the ignorance of the gardeners *"for they do not understand the culture and the grafting of trees as we do in France."* But he has praised the fruits grown in Kashmir and regards them of excellent quality.[206] Tavernier has referred to the cultivation of good quality of mangoes at Kolaras[207], Ahmadabad and Goa.[208] Other fruits of Goa mentioned by Tavernier are *ananas*, figs, plantains and coconuts.[209] In Maharashtra also we get references of the cultivation of the fruits like water-melons, coconut, jack-fruit, sugarcane, lemon, bananas, mangoes, *ananas* and cashew-nuts.[210]

In all parts of the Empire a wide variety of fruits and flowers were grown. According to Sujan Rai:

If one describes the multicoloured and well tasting fruits of the land, it will comprise of a separate book. Fruits can be had in both seasons, Rabi as well as in Kharif. Though grapes, melons (kharbuza), water-melons (tarbuz), pomegranates, apples, peaches and fig, etc. are better than vilayat, the specialty of land are kathal, budhal, ananas, kela, sharifa, coconut, knola, sangtara and kinar. These trees one can see in abundance.[211]

At Lahore melons and grapes grew in abundance and these were like those of Persia and Turkestan; its mangoes and oranges were better than Bengal and sugarcane was sweeter than that of Deccan.[212] Sujan Rai has greatly praised the *Nishkar* (sugarcane) and regarded it as the very good gift of God[213] Surat produced all types of fruits in large quantities, specially grapes and melons.[214] In Gujarat mango and other fruit trees were so numerous that the country was called an orchard. From Pattan (Anhilwara) to Baroda, a distance of 100 *Kos*, the area was full of mango trees, which yielded the finest fruits, some of which tasted sweet even when unripe. Other fruits grown here were figs, melons which were available both in summer as well as winter alike; grapes and roses were plentiful.[215] Sujan Rai has regarded the mango as the top fruit of the land

Its color is very attractive. It smells well and has very good taste. In sweetness it rivals sugar. Its tree is very attractive and shady.[216]

He has praised the mangoes of Bayana. Mangoes here sometimes reached the weight of one *seer* per piece.[217] Regarding the fruits grown at Thatta, Mohammad Saleh Kamboh writes:

In this part of world... fruits do not get spoiled for four months. Different types of fruits and flowers can be had every where.

Among the fruits grown there were mangoes which were very good and also small melons were grown here.[218] Baglana, a locality between Surat and the mountains, produced a variety of fruits viz. apricots, plums, apples, grapes, pomegranates, mangoes, *turanjes* and *ananas*.²¹⁹ It was regarded as second to Kashmir in fruit cultivation. The weight of a bunch of *fakhri* grapes was 8 *asars*²²⁰ oranges, were of the same weight, and a mango an *asar* and a half. The *kamud* orange of the region was famous as a rarity throughout the Deccan and its mangoes were well known to the nobles and the commons of the Deccan for their large size and length of their season.²²¹ Large oranges also grew well in Cooch Bihar.²²² Other parts of the Empire which produced good quality fruits were Chamba, which was known for its sweet and delicious fruits,²²³ Pakhli where apricots, peaches and walnuts were grown;²²⁴ Bhakkar²²⁵ and Punjab where the muskmelons could be had all the year round. Excellent mangoes and vines were also grown there.²²⁶ The water-melons of Agra were famous for their softness and sweetness. Musk-melons of Shahjahanabad and water-melon or cucumber of Akbarabad were celebrated for their deliciousness in comparison to the same fruits cultivated in other parts of India. Since these were grown in abundance, these were very cheap. Delhi imported cucumbers from Agra.²²⁷ In the period of the later Mughals also we get references to the availability of good quality fruits in different parts of the country because of the efforts made by the Provincial rulers. A variety of fruits were grown in Awadh. Those were mangoes, guavas, black berries, yellow berries, musk-melons, water-melons, cucumber, and betel leaves (pan) and so forth. Apart from this cultivation of sweet scented flowers like roses of various kinds, *bela*, jasmine and many other were grown²²⁸ During the reign of Muhammad Shah there are references of plantation of mango groves at Bagpur on the Jamuna, 16 miles north of Hasanpur.²²⁹ In most parts of Burhanpur a large number of orchards were present. Special varieties of musk-melons called *Firangi kharbuza* and *kharbuza-i-Hindustani* grew there in a large quantities. And many varieties of *kelas* (bananas) were also available.²³⁰ Patna was equally fortunate in respect of fruits, quality pan (betel leaf), jack-fruits, occurred in abundance but the most plentiful fruit grown in the region was mango. Its orchards and groves were present in the south of the city.²³¹ In *Sarkar* Hajipur (at the confluence of Ganga and Gandak, was about 8 *miles* from Patna)²³² fruits were grown "*in great* *plenty*". It was noted for its abundant growth of *Kathal* (jack-fruit) and *barhal*. The former then used to *"attain such a size that a man can with difficulty carry one"*, sometimes a yard long and half a yard broad. Grapes were also grown in there. In Sarkar Trihut delightful groves of orange trees extended to a distance of 30 miles. *Tut*, a sort of "mulberry" about 3 inches long and about as thick at the great end as a man's *"little finger"*, and *"a very sweet and pleasant fruit"* was grown in Shah Shuja's garden at Rajmahal. Baghalpur and its environs had abundance of *toddy* (*tar*) trees and gardens of mango trees. In Barari there was a "good garden of mangoes, trees were set all in rows in squares very handsomely"[233] From the *Akhbarat* of 1751– 52, reign of Ahmad Shah, there are references that different varieties of fruits especially mangoes and dry fruits constituted an important item of the diet of royal family.[234] In later period Britishers also cultivated a different varieties of fruits and flowers in their gardens. In their gardens were grown flowers like jasmine (for beauty and delight), and fruits like coconuts, guavas, mangoes, the delight of India, plum, pomegranates and bananas; a sort of plantain.[235] During this time India produced mangoes, pine-apples, plantains, pomegranates, pumplenoses, jacks, custard-apples, leeches, guavas, melons, oranges, lemons, limes, grapes, soursops, almonds, gooseberries, strawberries, tamarinds, plums, figs dates, citrons, loquats, most of these fruits were found on the tables of Europeans.[236]

Although the fruit cultivation was intensified on a large scale and almost every variety of fruit was grown in every part of empire. But there were certain fruits which were peculiar to certain regions. These regions were famous for different varieties of fruits known for their quality, taste, flavour, etc. The most favourite and the most common fruit of India was mango. It was typically an Indian fruit. There were many varieties of it. Although mangoes were found every where but those of Bengal, Gujarat,[237] Malwah, Khandesh and the Deccan were very good. In these regions its cultivation was encouraged by Akbar.[238] During the season a constant supply to the imperial fruit-house was ensured from Agra,[239] Hasilpur,[240] the Deccan, Burhanpur, Gujarat and the Malwah. Mangoes of Burhanpur were so big that a mango weighted fifty two and a half *tolas*.[241] Bernier has regarded the mangoes of Bengal, Golconda and Goa of excellent quality.[242] Sugarcane was grown chiefly in Agra (Bayana and Kalpi), Allahabad, Awadh, Delhi (*Mahamin Sarkar* Hisar Firoza), Lahore, Multan, Sind, Gujarat, Malwah and Ajmer.[243] De laet writes that the whole country from Agra to Lahore was *"cultivated and produced abundance of sugarcane."* Ajmer used to grow excellent *paunda* and fetched the highest price. During the 17th century the cultivation of sugarcane was extended to other places like kotah in Rajputana, Khandesh, Malwah, Bundelkhand, some places in Bihar, Bengal and Assam as well as Berar, Baglana, Aurangabad and Konkan. The black canes of Aurangabad province were so juicy that each yielded 5 *seers* of juice.[244] Sylhet was famous for sweet oranges *(santarah)*. These were very large and sweet.[245] Hajipur (in Bihar) was famous for jack-fruits *(katahal)* and *Baryhal* (a small round fruit). The former used to attain such a size that a man can with difficulty carry one.[246] Trihut (in Bihar) was famous for oranges. Groves of orange trees extended to a distance of 30 kos.[247] Illahabad (Allahabad) and Agra were known for grapes and melons,[248] fine pomegranates were grown in Satgaon and Hugli.[249] Jalalabad and Thatta were also famous for the cultivation of good quality pomegranates. Inyat Khan in his *The ShahJahan-nama* writes:

... otherwise pomegranates are not usually good in the imperial dominions expect at Jalalabad and Thatta...[250]

Bengal produced a fruit called kola that resembled orange but was better than orange. Another fruit called latkan, which was equal to gardgani also grew in Bengal.[251] Bengal plantains *(i.e. chini-champa)* were also famous for their sweetness.[252] The kamud orange of Baghlana was famous as a rarity through out the Deccan and the mangoes of Baghlana were well known for their large size and length of the season.[253] Kashmir and Kabul were famous for their cherries and apricots. Apricots of Kabul were so good that they could not be compared to those of other regions.[254] Kabul was also known for its excellent grapes especially the varieties of the *sahibi and kishmishi*.[255] Istalif in Kabul was famous for its peaches. Jahangir has praised them very much. One peach weighed about 63 *Akbari* rupees or 60 *tolas*. These peaches were sweet in taste.[256] Ahmadnagar was well known for its *fakhri* grapes.[257] Burhanpur was also known for its figs which were usually large and sweet.[258] *Bangash* apples were so good in sweetness, delicacy and taste that there was no comparison.[259] The Kashmiri apples were also renowned for their quality.[260]

With the efforts of the Mughals Horticulture received an impetus and fruit cultivation was intensified on a large scale. Earlier in the ancient and sultanate period a particular fruit was confined to only one or two regions but during Mughal period its extent was increased. It was extended to every part of the empire like Kashmir, Lahore, Delhi, Agra, Allahabad, Awadh, Bengal, Bihar, Malwa, Gujarat, Multan, Ajmer, Khandesh, Deccan and many other areas. Fruit cultivation was done in groves, orchards, gardens owned by emperors, by the rich nobles or officials, and by the peasants themselves and let out to cultivators and fruit sellers. Many central Asian fruits were introduced in India by the technique of grafting. Earlier cherries were not grown in Kashmir, but now they were introduced there and also by this method the quality of various fruits was improved. Not only this new species of fruits were introduced from the New World such papaya, cashew-nut and many more. As a result almost every sort of fruit was available in the country. In addition to this the fruit orchards and trees which were planted on the road sides not only lent beauty to the towns, but provided shades from the sun and a cool retreat for those seeking comfortable solitude. It also led to an increase in the availability of fruits to the travellers. With the result the facility on the road increased. It also led to the popularity of the trees. Persons of different regions acquired the knowledge of different plants, fruits and trees and also made an effort to grow these in their respective areas, which led to an increased availability of fruits throughout the empire.

REFERENCES AND NOTES

1. Randhawa, Mohinder Singh. *Gardens through the Ages*. Madras:The Mac Millan Company of India Ltd, 1976. p-46, hereafter Randhawa; *Gardens through the Ages*.
2. Basham, A.L. *The wonder that was India*. Calcutta: Rupa & co., 1996. pp-202, 203. Hereafter Basham; *The Wonder that was India*.
3. Randhawa, Op.cit. pp-68,69.
3a. Beal, Samuel. *SI-YU-KI, Buddhist records of the Western World*. Translated from the Chinese of Hiuen Tsiang AD 629. rpt. New Delhi: Munshiram Manoharlal Publishers Pvt. Ltd.,2014. p-88
4. Randhawa, *Gardens through the Ages*. p-54.
5. Pant, D. *The Commercial Policy of the Moguls*. Delhi: Idarah-i Adabiyat-i Delli, 1978. p-88. Hereafter Pant, *The Commercial Policy*.
6. Ansari, M.A. *Geographical Glimpses of Medieval India* Vol.I. Intro. With English Abstracts Jaweed Ashraf, Tasneem Ahmad. Delhi: Idarah-i-Adabiyat-i, Delli,1989. p-intro.[xx]. Hereafter Ansari, *Geographical Glimpses* vol.I
7. Afif, Shams Siraj. Medieval in Transition. Tarikh-i-Firozshahi. R.C.Jauhari ed. New Delhi: Sandeep Prakashan, 2001. pp-91-93
8. Ibid. pp-91-93
9. Ansari; *Geographical Glimpses*, Vol.2; p-intro.[xvii].
10. Firoz Tughluq a king noted for his many works of public utility.
11. Father Monserrate, S.J. *The Commentary of Father Monserrate, S.J.- on his journey to the court of Akbar*. trans. from Latin by J.S.Hoyland, Annotated by S.N.Banerjee. New Delhi: Asian Educational Services, 1992. p-96. Hereafter Father Monserrate.
12. The *Tarik-i-Rashidi* of Mirza Muhammad Haider Dughlat, Trans. by E Denison, Delhi: ABI publishers,1986. pp 428, 429. Hereafter *Tarikh-i-Rashidi*.
13. Ashraf, K.M. *Life and Conditions of the People of Hindustan*. third ed., New Delhi: Munshiram Manohar Lal Publishers, 1988. pp-119, 120. Hereafter K.M.Ashraf, *Life and Conditions*.
14. The *Rahla* of Ibn Batuta in M.A.Ansari's Geographical Glimpses vol.1.pp-40, 41. Also see The Travels of Ibn Batuta Vol.3. London: 1971; pp-609-11.
15. K.M.Ashraf. *Life and Conditions*, p-120.
15a. Babur, Zahiruddin Mohammad. *Baburnama*. Volumes I and II in one format. Trans. by A. Beveridge. reprint, Delhi: Low Price Publications, 2000. p-484 Hereafter *Baburnama*
16. Jena, Dr.Krusnachandra. *Baburnama and Babur*. Delhi: 1978. p-17. Hereafter Jena. *Baburnama and Babur*

17. Gulbadan Begam's *Humayunnama*. Trans.by Annette S.Beveridge; First Indian edition, Delhi: Oriental Books Corporation, 1983. p-22. Hereafter Gulbadan's Humayunama.
18. *Baburnama*. p-77.
19. Gulbadan's *Humauyn-nama*, p-8.
20. Allami, Abul Fazl. *The Ain-i-Akbari* vol.I. Transr. by H.Blochmann; reprint Delhi: Low Price Publications, 1997. P-68. Hereafter *Ain* Vol.I
21. Early, Abraham. *The Last Spring, Lives and Times of the Great Mughals*. first edition; New Delhi: Viking India,1997. p-249. Hereafter Early, *The Last Spring*.
22. Manucci, Niccolao. *Storia Do Mogor or Mogul India (1653 – 1708)* Vol.1. Trans. by William Irvine. Calcutta: Editions Indian,1965. p-199. Hereafter Manucci.
23. Bernier, Francois. *Travels in the Mogul Empire*.Trans.by A.Constable. New Delhi: Atlantic Publishers,1989. p-249. Hereafter Bernier.
24. Faruqi, Zahiruddin. *Aurangzeb and His Times*. Delhi: Idarah-i Adabiyat-i Delli,1980. p-508.
25. Zaidpuri, Ghulam Hussain's *Riyaz-us-Salatin* in M.A.Ansari's Geographical Glimpses. Vol.3, p-58.
26. Muhammad Umar. *Urban Culture in Northern India* during the 18th Century. Aligarh: 2001. p-241. Hereafter Umar, *Urban Culture*.
27. *Baburnama*, p-518.
28. *Ibid.*, pp-503-513.
29. Yadgar, Ahmad's *Tarikh-i-Salatin-i-Afghana* in the *History of India as told by its own Historians*, by Elliot and Dowson, vol.5. First edition Delhi:Kitab Mahal (WD) Private Ltd, 1964. p-24.
30. *Baburnama*, pp-502-513.
31. *Ibid.*, pp-202, 203.
32. A large green fruit shaped something like a citron; also a large sort of cucumber. *Baburnama*, p-203.
33. *Ibid.* p-203.
34. *Ibid.* pp-503, 504.
35. *Ibid.* p-504.
36. *Ibid.* pp504-513.
37. *Tarikh-i-Rashidi*. pp-425, 430.
38. *Ain* Vol.1. p-70.
39. *Ibid.* p-71.
40. Ansari. *Geographical Glimpses*, Vol.2. p-intro.[xv].
40a. Crowe, Sylvia and Sheila Haywood. *The Gardens of the Great Mughals*. Delhi: Vikas Publishing House Pvt. Ltd., 1973. p-62
41. Jena, *Baburnama and Babur*; p-18.

42. Aziz, Matto Abdul. *Kashmir under the Mughals [1586-1752].* Srinagar: Humayun Publishing House, 1988. p-202.
43. Ellison. *Nurjahan,* p-247.
43a. Terry, Edward. A Voyage to East India. Reprint, London: Gale Ecco Print Editions,1777. pp-90-91
44. Della Valle, Pietro. *The Travels of Pietro Della Valle in India* Vols. I & II ed. by Edward Grey. New Delhi:Asian Educational services, 1991.p-115. Hereafter Della Valle.
45. Aziz, Mattoo. *Kashmir under the Mughals.* P-202.
46. Foster, William.ed. *Early Travels in India (1583-1619).* London: Oxford University Press, 1921. p-158.
47. Manucci Vol.2; p-174.
48. Sarkar, Jadunath, *Fall of the Mughal Empire.* Vol.1. reprint New Delhi: Orient Black swan Pvt.Ltd.,1997. p-6.
49. Umar. *Urban Culture-18th Century,* p-241.
50. *Baburnama,* pp-686, 203, 208.
51. Jahangir. *The Jahangirnama* - memoirs of Jahangir, Emperor of India. Trans. Wheeler M. Thackston, New York: Oxford university Press, 1999 p-24. Also see *Tuzuk* Vol.1, p-5.
52. Randhawa, *Gardens Through the Ages*; p-112.
53. Allami, Abul Fazl. *The Ain-i-Akbari.* Volumes II and III, Trans. by Col. H.S. Jarret. reprint Delhi: Low Price Publications,1994. Vol. III, p-10.
54. *Ain.* Vol.II, pp-316-17.
55. The *Jahangirnama,* p-74.
56. The *Jahangirnama,* pp-247, 248. Also see *Tuzuk,* Vol.1, p-432.
57. Jahangir. *Tuzuk-i-Jahangiri* or *Memoirs of Jahangir.*Trans.by Alexander Rogers. Volumes I & II compiled in one. reprint Delhi: Low Price Publications, 1999. Vol.2, p-159.
58. The *Jahangirnama,* p-333.
59. *Ibid.* p-333.
60. *Paywandi* means to Graft; *ibid;* p-81. Also see *Tuzuk,* p-116.
61. Habib, Irfan. *The Agrarian System of Mughal India.* 2nd revised edition, New Delhi: Oxford University Press,1999. pp-55-56.
62. The *Jahangirnama.* -197.
63. Irfan Habib. *Agrarian System,* pp-53, 54.
64. The *Jahangirnama,* p-296.
65. Ellison. *Nurjahan,* -247.
66. Randhawa; *Gardens through the Ages.* P-122.
67. Qureshi, I.H. *The Administration of Mughal Empire.* Reprint, Delhi: Low Price Publications,1994. P-174. Also see Urbanisation and Urban Centres under the Great Mughals, p-43.

68. Faruki. *Aurangzeb and His Times*; p-16.
69. Irfan Habib. *Agrarian System*; p-56.
70. Manucci Vol.3, pp-170-171.
71. Bernier, p-249.
72. Darbari, Neera. *Northern India under Aurangzeb (social and economic conditions)*. Meerut: Pragati Prakashan,1982.pp-58-59.
73. Bhandari, Sujan Rai. *Khulasat-ut-Tawarikh*, Zafar Hasan ed. Trans.by J.N.Sarkar, Delhi: J&Sons,1918. p-110. Hereafter *Khulasat*.
74. Ansari, *Geographical Glimpses*; Vol.3, p-58.
75. Sarkar. *Fall of the Mughal Empire*, Vol.1, pp-4,6.
76. *Ibid*. p-195.
77. Randhawa. *Gardens Through the Ages*, p-143.
78. *Ibid.* p-145.
79. Newton, Denis Kincaid. *British Social Life in India*. First edition Newton Abbot:Readers Union,1974 p-62.
80. Bayly, C.A. *Rulers, Townsmen and Bazaars North Indian Society, In the Age of British Expansion*. New Delhi: Oxford University Press,1993 p-43. Hereafter C.A.Bayly.
81. *Ibid*. p-216.
82. Naqvi, H.K. *Urbanisation and Urban Centres Under the Great Mughals*. Shimla: Indian Institute of Advanced Study, 1972 p-43. Hereafter Naqvi; *Urbanisation and Urban Centres*.
83. Irfan Habib. *Agrarian System*, p-55.
84. Ain vol.I, p-73. Also see Vol.II, p-257
85. By Franks' ports he means primarily Camby, Surat and Goa, where Portuguese had established themselves. *The Jahangirnama*, p-25, f.n.-15
86. *Ibid*, p-24.
87. Kamboh, Mohammad Saleh. *Amal-i-Saleh* of in *Geographical Glimpses*; Vol.3; p-19.
88. Kulkarni, Anant Ramchandra. *Maharashtra in the Age of Shivaji*. New Delhi: Diamond Publications,2008 p-118
89 Bernier, p-442.
90. Qureshi. *The Administration of Mughal Empire*, p-174.
91. Irfan Habib. *Agrarian System*, p-55.
92. Della Valle Vol.1; p-134.
93. Manucci vol.2, p-158.
94. Ansari. *Geographical Glimpses* Vol.3, p-[xxii] intr.
95. Qureshi. *The Administration of the Mughal Empire*; p-171.
96. *Ain* Vol.2, p-72.
97. Moreland, W.H. *The Agrarian System of the Moslem India*. New Delhi: Atlantic Publishers,1994.p-127.

98. *Tuzuk,* Vol.2, p-52.
99. Irfan Habib. *Agrarian System,* p-285.
100. Sarkar, *J.N. Mughal Administration.* Patna: Superintendent, government Printing, Bihar and Orissa, 1920. p-173.
101. Ansari, M.A. *Social Life of Mughal Emperors (1526-1707)* New Delhi: Geetanjali Publishing House,1974. p-34.
102. *Ain* Vol.1, p-69.
103 *Tuzuk,* Vol.1, p-5.
104. *Ibid.* p-423.
105. *Ibid.* p-435.
106. *Ibid.* p-423.
107. *Tuzuk,* Vol.2, p-101.
108. *Tuzuk,* Vol.1, p-423.
109. Naqvi. *Urbanisation and Urban Centres,* Vol.1, p-104.
110. Ansari, *Geographical Glimpses* Vol.3, p-50.
111. Sarkar. *Fall of the Mughal Empire* Vol.1, p-10.
112. Naqvi. *Urbanisation and Urban Centre,* p-44.
113. *Baburnama,* pp-514, 515.
114. *Ibid.* p-610.
115. *Ain* Vol.1, p-93.
116. *Ibid.* pp-81,82,93.
117. *Ibid.* p-81.
118. *Ibid.* p-82.
119. *Ibid.* p-93.
120. *Ain* Vol.2, p-352.
121. Tuzuk Vol.1, p-5. Also see *The Jahangirnama,* p-24.
122. *Ibid.* pp-5,6.
123. *The Jahangirnama,* p-270.
124. *Ibid.* p-332.
125. *Ibid.* p-433.
126. Finch, William. *India as seen by William Finch.* edited by R.Nath Jaipur: Historical Resaerch Documentation Programme, May 1995.p-79.
127. Della Valle Vol.1, p-115.
128. Irfan Habib. *Agrarian System,* p-53.
129. *The Jahangirnama,* p-249. Also see *Latifa-i-Faiyazi* by Nooruddin in *Geographical Glimpses,* Vol.2, p-24.
130. Khan, Inayat. *The Shahjahannama.* Trans. W.E. Begley and Z.A. Desai. New Delhi: Oxford University Press,1990. p-125.
131. Ellison. *Nurjahan,* p-247.
132. *Amal-i-Saleh* by Mohammad Saleh in *Geographical Glimpses;* Vol.3, p-13.
133. Manucci. Vol.2, pp-315, 316.

134. *Khulasat*, p-120.
135. *Khulasat in Geographical Glimpses*; Vol.3, p-63.
136. Francois, Bernier, *Aurangzeb in Kashmir (Travels in the Mogul Empire)*.Trans. Irving Brook, edited by D.C.Sharma. Delhi: Rima Publishing House, 1988.p-66.
137. *Amal-i-Saleh in Geographical Glimpses* Vol.3, pp-14, 15, 16, 19.
138. *Masalik-ul-Absar-Fi-Mumalik-ul-Amsar* by Shahabuddin-Al-Umar in *Geographical Glimpses* Vol.1, p-31.
139. Dayal, Maheshwar. *Rediscovering Delhi*, New Delhi: S.Chand & Co. pvt. ltd.,1975. p-231.
140. Stocqueler, J.H. *India under the British Empire*. Delhi: Concept Publishing House, 1992. p-100.
141. *British Social Life in India*; p-62.
142. Naqvi; *Urbanisation and Urban Centres*; pp-67, 68.
143. Haig, Sir Wholseley. *The Cambridge History of India*, Vol.4, *The Mughal Period*, edited by Sir Richard Burn. Delhi: Cambridge University Press, 1979. p-57; Also See D.Pant; The Commercial Policy; p-54; H.G.Trevaskis, An Economic History of Punjab; p-110.
144. Farooque. *Roads and Communications in Mughal India*. Delhi: Idarah-i Adabiyat-i Delli, 1977. p-55.
145. Ansari, M.A. *European Travelers under the Mughals*. Delhi: Idarah-i Adabiyat-i Delli,1975. p-3.
146. *Tuzuk* Vol.1, p-107.
147. Farooque. *Roads and Communications*; intro p-xxii.
148. *Tuzuk* Vol.2, p-100.
149. *India as seen by William Finch*, p-72.
150. De Laet. *The Empire of the Great Mogol (De Imperio Magni Mogolis).De Laet's description of India*. Trans. by J.S.Hoyland, NewDelhi: Munshiram Manohar Lal Publishers, 1974. p-54. Also see Chetan Singh, p-206.
151. Naqvi. *Urbanisation and Urban Centres*; p-98.
152. *Early Travels*, p-244. Also see Ansari. *European Travellers*; p-60.
153. *Tuzuk* Vol.2, p-100.
154. Manucci Vol.1, p -159.
155. Travernier Vol.1, p-78.
156. Pant, D. *The Commercial Policy of the Moguls*. rpt.Delhi: Idarah-i Adabiyat-i Delli,1978.p-129.
157. The Bara Pala Bridge leading to Humayun's Tomb. D Laet, p-47.
158. De Laet, pp-47, 48.
159. *Early Travels*; p-156.
160. De Laet, p-91.
161. Check Bernier; Also see H.G.Keene, Handbook to Agra; pp-21-22.

162. *Aurangzeb in Kashmir*; p-67.
163. Sarkar. *Fall of the Mughal Empire,* Vol.1, p-6.
164. Anonymous; *"Military Defence of Our Empire in the East"*. Calcutta review vol.II, Cambridge University Press,1844 p-36. Also see Farooque, Roads and Communications, p-54.
165. *Ain* Vol.1, p-68.
166. Father Monserrate; p-213.
167. *Ain* Vol.2, p-354.
168. Naqvi. *Urabisation and Urban Centres,* p-24.
169. Qazvini, Mohammad Amin. *Badshahnama* in *Geographical Glimpses.* Vol.3, pp-2,3.
170. *Ain* Vol.1, p-68.
171. *Amal-i-Saleh in Geographical Glimpses* Vol.3. p-9.
172. *Ain* Vol.1, pp-68, 69.
173. Father Monserrate, p-104.
174. *Ain* Vol.1, pp-69,70.
175. *Ibid,* p-71.
176. *Tuzuk* Vol.1, p-5.
177. Pelsaert, Francisco. *Jahangir's India, the Remonstrantee of the Francisco Pelsaert.* Trans. W.H.Moreland and P.Geyl. Delhi: Idarah-i Adabiyat-i Delli, 1972. p-48.
178. *Ibid,* p-35.
179. *Ain* Vol.1, pp-70, 71.
180. De Laet, pp-43,44.
181. Ansari. *European Travelers,* p-35.
182. *Early Travels,* p-297.
183. *Ibid,* p-166.
184. Della Valle Vol.1, p-47.
185. Manucci Vol.3, pp-171-173.
186. *Tuzuk* Vol.1, p-116. Also see *The Jahangirnama,* p-81.
187. *The Jahangirnama,* p-82.
188. Manucci vol.2, p-174.
189. *Tuzuk* Vol-1, p-5. Also see *The Jahangirnama,* p-24.
190. *The Jahangirnama,* p-95.
191. *Ibid,* p-340.
192. Sarkar, Jagdish Narayan. *Studies in Economic Life in Mughal India.* Delhi: Oriental publishers and distributors, 1975 p-238.
193. *The Shahjahannama* of Inyat Khan, p-127.
194. *Ibid,* p-127.
195. *Chahar Chaman* by Chandra Bhan Brahman, *Geographical Glimpses* Vol.3, p-6.
196. Bernier, p-249.

197. Ibid, p-249.
198. Ibid, p-249.
199. Sarkar. *Mughal Administration*, p-64.
200. Bernier, p-249.
201. Ibid, p-250.
202. Ibid, p-397.
203. *Khulasat*, pp-120, 121.
204. Bernier, p-438; also see *Riyaz-us-Salatin* of Ghulam Hussain in Geographical Glimpses Vol.3, p.58.
205. *Riyaz-us-Salatin* in *Geographical Glimpses* Vol.3, p-58.
206. Bernier, p-397.
207. Kolaras, a well known town in Gwalior, though not mentioned in the Imperial Gazetteer. The total distance from Mughal Sarai to Kolaras measured on the map is about 62 miles. Tavernier Vol.1, p-48.
208. Ibid, pp-48, 65, 150.
209. Ibid, p-150.
210. *Maharastra in the Age of Shivaji*, p-118.
211. *Khulasat* in *Geographical Glimpses* Vol.3, p-62.
212. *Khulasat*, p-110.
213. *Ibid*, p-62.
214. *Amal-i-Saleh* in *Geographical Glimpses* Vol.3, p.19.
215. *Khulasat*; p-66. Also see *Amal-i-Saleh* in *Geographical Glimpses* Vol.3, and *Mirat-i-Ahmadi* in Geographical Glimpses, Vol.3, p-38.
216. *Khulasat* in *Geographical Glimpses* Vol.3 p-62.
217. *Ibid*, p-69.
218. *Amal-i-Saleh*, Geographical Glimpses Vol.3, p-21, also see *Khulasat*, p-68.
219. *Ibid*, p-19, also see *Aurangzeb in Muntakhab-al-lubab;* p-21.
220. An *asar* was equal to a *seer*, and a *seer* in the time of Shahjahan was reckoned as the weight of 40 dams or copper coins. *Ibid.p*-21.
221. *Ibid*, p-21.
222. *Riyaz-us-Salatin in Geographical Glimpses*, Vol.3, p-53.
223 *Khulasat*, p-104.
224 *Ibid*, p-120.
225 *Ibid*, p-70.
226. *Ibid*, intro. ixxvi.
227. Muhammad Umar, *Urban Culture*, p-267.
228. Dr.S.Moinul Haq, Journal of Pakistan Historical Society (Industries and Commerce – Qidwai in the Kingdom of Awadh). p-309.
229. Sir Wholsely Haiq, *The Cambridge History of India* Vol.IV. *The Mughal Period*, pp-345, 346.
230. *Latifa-i-Faiyazi* of Nooruddin in *Geographical Glimpses*, Vol.2, p-24.

231. Naqvi, H.K. *Urban Centres and Industries in Upper India (1556-1803)*. Bombay: Asia Publishing House,1968. pp-93,94. Hereafter Naqvi, Upper India (1556-1803).
232. Sarkar. *Studies in Economic Life in Mughal India*, p-237.
233. *Ibid*, p-296.
234. Verma, B.D.(ed.) *Akhbarat, News letters of the Mughal Court, Reign of Ahmed Shah, 1751-52*. Bombay,1949. p-2.
235. *British Social Life in India*, p-62.
236. *India under the British Empire*, pp-111, 112.
237. *Baburnama*, p-504.
238. *Ain*.Vol.1 P-72.
239. *Tuzuk*, Vol.1, p-5.
240. *Ibid*, p-362.
241. *The Jahangirnama*, pp-220, 95.
242. Bernier, p-249.
243. *Ain*.Vol-2, pp-77, 80, 83, 85, 88, 93. Also see *Mughal Economy*, pp-17,18 and Naqvi, *Urbanisation and Urban Centres*, Vol.1, p-24.
244. Sarkar, *Mughal Economy*, P-18.
245. *Ain*, Vol-2, p-136.
246. *Ibid*, p-164.
247. *Ibid*, p-165.
248. *Ibid*, pp-169, 190.
249. *Ibid*, p-137.
250. *The Shahjahan-nama* of Inyat Khan, p-395. Also see Father Monserrate. P-149. *The Jahangirnama*, pp-226, 227.
251. *Haft Iqlim* by Ami Ahmad Razi in *Geographical Glimpses*, Vol.2, p-22.
252. *Baburnama*, pp-504, 505.
253. Khan, Khafi. *Muntakhab-al-lubab*. Trans. Anees Jahan. Bombay: Somaiya publications, 1977. *p-21*.
254. *The Jahangirnama*, pp-74, 81, 333.
255. *Ibid*, p-81. Also see *Ain* Vol.1, p-69.
256. *The Jahangirnama*, p-82.
257. *Ibid*, p-211.
258. *Ibid*, p-239.
259. *Ibid*; p-310.
260. *Ibid*; p-333.

ROYAL GARDENS

The history of gardens is closely related with the history of civilization. With the development of civilization there started man's quest in every aspect of nature, whether it was study of stars or study of plants. With the passage of time man discovered various things which included arithmetic, Astronomy, medical science and also patterns of gardening. The history of garden and gardening is also connected with the history of people and their culture, which includes their science, art, literature, as well as religion. And gardening can also not be isolated from agriculture because horticulture is also a branch of agriculture and the knowledge which the man acquired in growing crops was also applicable in growing of ornamental plants and fruits trees.[1] In the holy books there is frequent mention of gardens for instance *"Garden of Eden"* in Bible. Paradise in Bible has always been visualised as full of flowering and fruit trees which provides harmony and soothing affect to man. In Quran also garden is constantly cited as a symbol for paradise with shade and water as its ideal elements. *"Gardens underneath which river flow"* is a frequently used expression for the bliss of the faithful, and occurs more than thirty times throughout the Quran.[2] In Quran it is referred that God has actually defined paradise as a garden and it is upto the individual not only to aspire to it in the after life, but also try to create its image i.e. something like it on earth.

Ornamental gardening and landscape gardening are ancient arts. We find the references of gardens in the ancient civilizations of the world. Mesopotamia, the birth place of agriculture, was also the cradle of gardening. The Mesopotamians constructed private parks and terraced gardens – usually on artificial mounds or supported by columns as the Hanging Gardens of Babylon. The Egyptians built formal walled gardens.[3] The Egyptians had mastered the art of canal irrigation. The large gardens of Pharaohs and nobles were irrigated by canals which distributed the water of Nile. The small gardens were irrigated by the *Shadoof*.[4] Ancient Indians also greatly loved flowers and trees. In the epics like *Ramayan* and *Mahabharata*, it has been laid down that to the folk of that time forest was a realm of mystery. It was inhabited by scholars and anchorites. It was full of beautiful flowers and fragrances. It was a haunt of sweet singing birds; and it was cool and green. All holiness might be attained under its soothing influence.[5] During Mauryan period (322 BC-187 BC) there are references to the construction of gardens. Megathenes in his *Indica* describes with wonder the beautiful parks surrounding the palace of Chander Gupta Maurya (322-298 BC), and many references in Sanskrit literature show that wealthy citizens had gardens attached to their houses and large parks in the suburbs containing partitions in which they spent much of their leisure time. In hot climates an expanse of water was an almost essential feature of the garden, the parks of wealthy contained artificial lakes and pools, often with fountains. Apart from private gardens, there were also public gardens. In the vicinity of most cities were the groves which were the favourite resorts of the town people. Ashoka (268-232 BC) took pride in the fact that he had planted such groves for the recreation of man and beast, and some other kings are recorded having followed his example.[6] In the *Meghaduta* of Kalidasa, *Udyans* i.e. gardens are frequently mentioned. Four kinds of gardens have been mentioned by Vatsayayana in his work *Kamasutra*, these are *Pramodudyan*, for the enjoyment of kings and queens, *Udyan*, where kings passed their time playing chess with their courtiers, *Brikshvatika*, where the ministers and courtiers made merry; and finally *Nandanvan*, dedicated to Lord Indra.[7]

Gardens were laid out in more sophisticated form Under the Delhi Sultanate (1206-1526 AD) than earlier. The orchards and gardens of fruits and fragrant flowers was a characteristic feature of any Muslim urban settlement.[8] During the Sultanate period the state as well as society encouraged the plantation of walled as well as unwalled gardens and orchards on a large scale. The contemporary sources mention a circle of thousand big gardens around almost all the cities. Towns had the same pattern, villages had their share of gardens both large and small.[9] Sultan Firoz Tugluq (1351–1388 AD) finds a special mention in the construction of gardens. He was very much fond of gardens. Due to his efforts 1022 gardens were laid in the villages around Delhi. Those who constructed gardens were given permanency on their land without any conditions or enquiries.[10] In the reign of Alaud'din Khilji (1296-1316 AD) 53 gardens were added to this list. Band Salora had 80 while Chittor[11] had 44 gardens.[12] Under the Lodis (1451-1526), the tomb gardens were constructed. The Lodi garden in Delhi is a fine example of this.[13] During this period Rajputana not only maintained but extended the tradition of laying out gardens. Apart from Chittor, Dholpur, Gwalior and Jodhpur also took up improved methods of cultivation and gardening. In Dholpur especially gardens shaded the route to the city for a distance of 7 *Krohs* [about 14 miles]. The art of gardening was understood in India long before the coming of Mughals.[14] Provincial rulers also made great contribution in the construction of gardens, for instance in *Tabqat-i-Akbari*, it is mentioned that Sultan Mahmud of Gujarat erected a special citadel (*Jahan Panah*) and laid out several gardens.[15] Certain inscriptions of Sultanate period also show that gardens were laid by the rulers for instance there is an inscription dated 6th October 1461 to 25th Sept.1462, on a tomb by Kamal Maula, it reads:

this is of garden of paradise of ... Qutub Kamal ... during his reign Mahmud Shah Khilji ... established them a new in 861.

So according to this inscription there was a garden of Qutub Kamal which was renovated by Sultan Mahmud Shah Khilji.[16] Although a number of gardens were laid out in India prior to the coming of Mughals but their construction was neither well planned nor intensified a large scale as during the Mughal period.

Gardens were among the most substantial features of Mughal architecture. The contemporary sources mention the laying of the gardens by Mughal emperors and their family members and governors throughout the Mughal period. The extent of these gardens spread from Delhi to Kabul and from Delhi to Gujarat. Although this period witnessed constant wars and revolts, but the making of gardens was a ruling passion. No account of the building art of the Mughals would be complete without a reference to the landscape architecture of this period, as illustrated by the large ornamental gardens which it was the pleasure of the rulers and others in power to lay out on certain appropriate sites. The idea of these retreats was brought from Persia, whose poets were far ever singing of their delight, as for instance, Firdausi, in describing the garden of Afrasiab, says

like the tapestry of the kings of Ormuz, the air is perfumed with musk, and the waters of the brooks are the essence of roses.[17]

The years between Babur's invasion in 1526 and the disposition of Shahjahan in 1658 roughly span the finest age of Mughal gardens. Mughals laid out numerous beautiful gardens and monuments. They laid out gardens – highly elegant, exquisite and tasteful throughout the empire, which served them as pleasure resorts or temporary quarters when they paid a visit to any particular locality. These gardens were constructed on a lavish scales with extensive grounds, causeways, small plots of flowers and fruits, water channels and residential quarters.[18] Not only the members of royal families but the nobles also contributed a lot in laying out of gardens. In some major cities of the empire – in Agra and Delhi, for example it was customary for the princes and nobles to construct residences and gardens where ever possible[19]

First Mughal Emperor Babur (1526 – 1540) was a great lover of nature and beauty Babur's love for nature and open life was inherited. His birth place Farghana was famous for beautiful gardens, orchards, fruits and flowers.[20] Babur's ancestors were extremely fond of laying out gardens. In *Baburnama*, he refers to various gardens laid by Timur Beg and Aulugh Beg[21] in the town and suburbs of Samarqand. Gardens laid out by Timur described in the *Baburnama* are *Bagh-i-bulandi* or garden of height, *Bagh-i-Dilkusha* or heart expanding garden, *Naqshil-jahan* (world's picture), *Bagh-i-chanar*, *Bagh-i-Shamel* and *Bagh-i-bihisht*.[22] *Bagh-i-Maidan* (garden of plain) was laid out by Aulugh Beg.[23] In fact the Timurides displayed a particularly strong love of gardens, perhaps in common with all Turkish people. Whenever Timur brought a new wife home, he had a special garden created for her.[23a] Babur himself was a great constructor of gardens. His Daughter Gulbadan Begam has described Babur as a builder, a lover of view, a maker of gardens and planter of trees.[24] Miniatures in *Baburnama* depict him overseeingg his gardeners as they prune, sow, plant tree striplings and carry corn seed in their shovels.[24a]

Babur laid out numerous gardens in Kabul. He writes in *Baburnama* that after he had taken Kabul in the year 1508-09, he laid out four gardens (*char bagh*) in Kabul known as *Bagh-i-wafa* (garden of fidelity).[25]

the garden lies high, has running water close at hand, and a mild winter climate. In the middle of it, a one-mill stream flows constantly past the little hill on which are the four gardens plots. In the south west part of it there is a reservoir, 10 by 10, round which are orange trees and a few pomegranates, the whole encircled by a trefoil meadow. This is the best part of the garden a most beautiful sight where oranges take colour. Truly that garden is admirably situated.[26]

Zain Khan in his *Tabqat-i-Baburi* has also given a beautiful description of this garden. He writes:

it is an excellent garden which causes heart sore to the garden of Iram.[27]

Besides, ten other gardens are mentioned in Amir Qazwin's *Padshannama*, which were Laid down by Babur in or near Kabul. Those were – the *Shahr-ara* (town adoring) which when Shah-i-Jahan first visited Kabul in the 12th year of his reign (1048 AH or 1638 AD) contained very fine plane trees Babur had planted. Other were the *Charbagh*, the *bagh-i-Jalankhana*, the *aurta bagh* (middle garden), the *Saurat bagh*, the *bagh-i-mahtab* (moon light garden), the *bagh-i-ahukhana* (garden of deer house), and three similar ones.[28] Babur was so much fond of these gardens that when he finally settled in India, he still found time to write to his governor in Kabul, Khavaja Kilan, with instructions that the gardens he had planted there should be kept well watered and properly maintained with flowers[28]

Since Babur belonged to the area where the gardening system was very much advanced, and accustomed to the luxurious gardens and buildings, he was very much disappointed to see the gardens of India. He writes in his memoirs:

the towns and country of Hindustan are greatly wanting in charm. Its towns and lands all are of one sort; there are no walls to the orchards (baghat), and most places are on dead level plain.[30]

He complains that there was no running water in the gardens of Hindustan.[31] Babur did not claim himself to be the innovator of gardens in India, he writes about the gardens of Bahlul Lodi (1451-1489) and Sikandar Lodi (1489-1517) and a garden of lemons in a valley near Gwalior.[32] But what Babur complained of was that gardens in Hindustan were not orderly and as symmetrical as those laid down by him and his nobles at Kabul. With Persian tradition of gardening and irrigation firmly embedded in his mind, Babur first thought of bringing the blessings of water to the hot and dusty land of Hind. Babur writes:

it kept coming to my mind that water should be made to flow by means of wheels erected where ever I might settle down, also that grounds should be laid out in an orderly and symmetrical way.[33]

Babur felt that things were not moving and progressing in India and people were still sticking to old methods and practices. Babur had some suggestions to make. He set examples by providing many things of cultural amenities. New kinds of large wells and *baolis* (stepped wells) were constructed; regularly planned and symmetrically arranged pleasure gardens with rose beds and narcissi were laid out.

Thus Babur also brought to it the love of orders and symmetry that was equally a part of the paradise garden tradition. According to Abul Fazl:

Formerly people used to plant their gardens without any order, but since the time of the arrival in India of the emperor Babur, a more methodical arrangements of the gardens has obtained, and travellers now a days admire the beauty of the palaces and the murmuring fountains.[34]

Babur laid out numerous gardens in India. Even in the midst of his campaigns he did not forget to lay out gardens at beautiful places. At least three gardens in Agra are attributed to Babur – *Aram Bagh, Dehra bagh* and *Zahara bagh*, the two latter being for his daughters. Babur's description of the first garden he laid out on the banks of Jamuna is a classic of the transformation of an unfavourable site. He writes:

... also that grounds should be laid out in an orderly and symmetrical way. With this object in view, we crossed the Jun-water to look at garden – grounds a few days after entering Agra. These grounds were so bad and unattractive that we traversed them with a hundred disgusts and repulsion. So ugly and displeasing were they, that the idea of making a charbagh in them passed from my mind, but the needs must! As there were no other land near Agra, that same ground was taken in hand a few days later. The beginning was made with a large well from which water comes for the hot bath, and also with the piece of ground where the tamarind trees and the octagonal tank now are. ... then in that charmless and disorderly Hind, plots of garden (baghcha) were seen laid out with order and symmetry, with suitable borders and parterres in every corner and in every border rose and narcissus in perfect arrangement.[35]

Zainkhan, author of *Tabqat-i-Baburi* has praised about this garden in this way:

In elegance and refinement it is like the terrestrial pleasure
In light and purity it is an object of envy for the sacred sanctuary of Mecca.
The spacious space is like the courtyard of the sublime Paradise,
Its refreshing wind is like the breaths of the faithful spirit (of angel Gabrial);
Sweet basil and fragrant hyacinth are in embrace with each other,
Rose flowers and jasmine are shoulder to shoulder (Shrugging their shoulder for joy)...[36]

The basic pattern of this garden was geometrically laid out walks, with platforms well above ground level from which to view the garden. The *Zahara bagh* was built for Babur's daughter Zahara. It is one of the largest paradise garden palaces in Agra, lying between the Arambagh and the site of *Chini-ka-Roza*.[37] About *Bagh-i-Gulafshan* Jahangir writes that Babur laid out this garden on the east of Jamuna:

a garden (charbagh) which few places equal in beauty. He gave it the name of Gulafshan (flower scatterer), and erected in it a small building of cut red stone....[38]

Babur laid out another garden about 6 miles from Agra and called it *Bagh-i-Zarafshan* (gold scattering garden).[39] Babur writes that he often visited this garden.[40] Gulbadan Begam also records Babur's visits to this garden.[41] Babur also constructed a garden at Sikri about its construction he writes:

leaving this place, we visited Biana[42] *again, went on to Sikri, dismounted there at the side of a garden which has been ordered made, stayed there for two days supervising the garden and on Thursday the 23rd of Rajab (April 25th) reached Agra.*[43]

Babur made a *Chaukandi*[44] in the Sikri garden and he used to put in it a *tur-khana*, where he used to sit and write his book.[45] Another garden was constructed by Babura at Kaldakahar, some 20 miles north of Bhira, in Kabul and called it *Bagh-i-Safa* (Garden of Purity).[46] Buildings of all kinds were included in the garden palaces of Babur, together with wells, reservoirs, aqueducts, and bath houses, the later being provided with hot water. One remarkable enterprise was a well containing a three storied house:

when the water is at its lowest, it is one step below the bottom chamber; when it rises in the rains, it sometimes goes into top storey.

Thus in the hottest weather, descent to the lowest floor could offer water and a cool atmosphere.[47] Till the end of his days Babur retained his love for nature and his aesthetic tastes. He had so love for gardens that he wished himself to be buried at his favourite garden at Kabul. When he died in 1530, he was first buried in Aram bagh at Agra. But in about 1544, in accordance to his will, which he had made before his death, his remains were taken to Kabul, probably by his Afghan wife Bibi Mubarika, in the terraced garden of his choice on the slope of the hill *Shah-i-Kabul*.[48]

Babur's successors cherished his memory, and were inspired by his enlightened out look his religious tolerance, his love of art and architecture and music and his interest in fruits and flowers and gardens. And his gardens became models for the subsequent Mughal gardens. Humayun (1530 – 1556) too was fond of gardens and from the contemporary sources it is evident that he preferred to stay in the gardens during his expeditions. Abul Fazl writes that while proceeding to Kashmir Humayun halted in a garden at ruhri near Bhakkar.[49] Gulbadan Begam too has described about Humayun's halt at this garden and she has written that this garden was made by Mirza Shah Hussain Samandar.[50] On his visit to Hirat, Humayun made an excursion to all palaces and gardens there.[51] Some other incidences also narrate Humayun's halt at gardens. After being defeated by Shershah, when Humayun's army approached Lahore, there Mirza Kamran came forward to meet him and did homage. At this place Humayun stayed in the most charming spot, the garden of Khwaja Dost Munshi, and his brother Hindal took up his quarters in the garden of Khwaja Ghazi, who was Mirza Kamran's Diwan.[52] Humayun himself made a significant contribution in construction of gardens. He was ably the first oriental Monarch to think of a floating garden. He ordered four large bazars to be set up in the river Jamuna and personally directed the changing of them into a floating city. His other ideas along with this experiment are given below:

In like manner the royal gardeners made, in accordance with orders, a garden on the river... "another of his inventions was a moveable bridge..." another of his inventions was a moveable palace.[53]

Khwandmir in *Qanun-i-Humayuni* writes that in the year 1531 Humayun got the floating gardens to be planted over river Jamuna.

And similarily the royal gardeners, according to the rders of this manifestation of Divine favour (the King), placed wooden planks on several of the boats, and spreading earth on them made them suitable for horticulture. They made fine orchards, on the boarders of which grew fruit trees and flowering plants, and all kinds of vegetables, tulips, jasmine were seen growing and flourishing in the river.

Verse:- Skill of clever persons made

   the moving gardens went round the world.[54]

Praising the beauty of this garden Khwandamir writes:

Verse:- (It was) as pleasing as the nature of intimate friends,
   And as beautiful as the faces of the witty companions (beloved persons)
   Certainly no eye saw such a garden,
   Neither in the spacious heavens nor on the wide earth.[54a]

Humayun also got many Gardens planted in the City of Dinpanah (Delhi). When Humayun planned the city of Dinpanah, he ordered that in this city a magnificent palace of seven storeys to be erected, surrounded by delightful gardens and orchards, of such elegance and beauty, that its fame might draw people from the remotest corners of the world for its inspection.[55] Father Monserrate writes that Humayun was devoted to architecture and had planted beautiful green trees and lovely parks at Delhi.[56] This period also witnessed the construction of the garden tombs. Humayun also completed one of the garden palaces of Babur at Agra. Humayun's tomb which was constructed after Humayun's death is one of the first garden tombs, of which the Mughal period produced so many splendid examples and is Humayun's great memorial. Humayun's widow Hamida Begam, supervised the construction of this tomb in Delhi, which was begun around 1560 and completed in 1573 A.D. A parterre rather than a garden, it takes the *charbagh* motive and enlarges and repeats it in an intricate pattern almost two dimensional and strongly Persian in character.[57] Father Monsserate who visited India during Akbar's reign has written:

this tomb is of a great size, and is surrounded by beautiful gardens.[58]

Afghan ruler Shershah Suri who succeeded Humayun by defeating in the battle of Kannuj (1540-1545) was also fond of laying out gardens. He planted trees on the Grand Trunk roads. Islam Shah (1545-1553), his son maintained the gardens made by his father. In the beginning of his reign, he issued orders that more *serais* should be build for the convenience of public and he also directed that the gardens which the Shershah had laid out should not be alienated, and that no changes should be made in their limits.[59]

Mughal emperor Akbar (1556 – 1605) contributed greatly in making of gardens. Two great projects of Akbar's time were the city of Fatehpur Sikri and the fort of Agra. The former a city of palaces, gardens, baths and tanks was built at the Shrine of Sheikh Salim.[60] In *Tarikh-i-Akbari* it has been written:

when emperor returned to Agra from Ajmer in the month of Rabi 2 A.H. 979 or August, 1571 A.,D., he issued an order for laying the foundation of Fatehpur Sikri. When the architects and planners presented the map of the proposed construction to the emperor, he instructed that houses to be built and gardens be planted within the compound and covering the neighbouring hills for about 2 to 3 Karohs.[61]

Many paradise resembling gardens were laid out, increasing the beauty of Imperial quarters. Jahangir too refers to plantation of the gardens in this city. He writes:

During the middle of the month of Rabi 979 [Aug.1571], an imperial decree was made for the foundation of a lofty fortress and delightful buildings to be raised ... and in a short time a magnificent city had come into being... all sorts of pleasure gardens added to the delight of the palace, and it was named Fatehpur.[62]

Akbar got planted magnificent gardens at Agra. He imported trees and flowers of every kind and got them planted in these gardens. Peter Mundy, an European traveller who visited one of the gardens writes:

the gardens about Agra are many, but the chiefest are Darre ca baug (Dehra bagh) of king Ecbars on this side of the river, and Moote Ca baug, on the other side, built by Noore Mohol... In some little groves are trees, as apple trees, (those scarse), or enge trees, mulberrie trees etts. In other squares are your flowers, herbs etts., where of roses, marigolds (theis scarse only in Moote ka baug) to be seen; French marigolds abundance; poppeas redd, carnation we knowe not in our parts, many growing on prettie trees, all watered by hands in tyme of drought, which is nine months in a year.[63]

Akbar followed the example of his forefathers and for his burial site he chose a garden on the road to Delhi, at 3 leagus distance from Agra on the West to which he gave the name of Sikandra.

This mausoleumis a very large dome… the garden is very large[64]

In Kashmir Akbar laid down numerous gardens. He visited Kashmir three times. He seems to have enjoyed the autumn colours and the saffron fields, and visited the spring of Achhabal, already a place of pilgrimage. He found a richly fertile country with abundant sun shine, all this made Akbar to call Kashmir his private garden. In fact Kashmir was the Mughal's ideal garden land. The journeys of the imperial court to Kashmir were celebrated, and at halting places numerous gardens were laid out. Wah *bagh* at Rajauri[65]; and many others like along the route, comparatively little visited today and some in ruins.[66] Wah *bagh* is one of the beautiful gardens of the Mughals. It is understood that the initiative of laying out this garden was taken by Akbar, he was so strucked with its beauty that it drew from him the exclamation of wah *bagh* (or what a garden).[67] But it was Jahangir who actually built the garden palace. But it was above all at Srinagar, the natural heart of Kashmir, the objectives of the journeys, that the finest gardens came into being. All round the Dal lake numerous gardens were laid out. A count taken in Jahangir's time showed over 700 of them. Akbar's own contribution are the fortress of Hari Parbat and the garden palace of Naseem bagh. Akbar's first building project in Srinagar was the fortress of *Hari Parbat*, which dominated Srinagar and lake Dal. He is said to have imported some 200 stone masons from India to build the fort, since masonry was little understood in Kashmir.[68] An account of this fortress and garden in it is also given by Jahangir in his *Tuzuk-i-Jahangiri*:

near the city there is a small hill which they call Kuh-i-maran, as well as Hari Parbat. My father gave an order that they should build in this place a very strong fort of stone and lime; this has been nearly completed during the reign of this suppliant, so that the little hill has been brought into the midst of the fortifications and the wall of the fort built round it…

In the palace there was a little garden, with a small building… At this period it appeared to me to be very much out of order and ruinous … I ordered Mu'tamidk, who is a servant ..., to make every effort to put the little garden in order and repair the building. In a short space of time... it acquired new beauty. In the garden he put up a lofty terrace 32 yards square, in three divisions and having repaired the building he adorned it with pictures by master hands, and so made it the envy of the picture gallery of China. I called this garden Nur-Afza (light increasing)[69]

The hill of *Hari Parbat* is linked by a canal to Dal lake, the head of the canal being marked by a *chinar* tree of immense dimensions, one of the largest in Kashmir. On the west bank of the lake lies Akbar's other legacy to Srinagar, the Naseem *bagh*. In fact it was the first great Mughal garden laid out in Kashmir in the year 1597. *Naseem bagh* the "garden of breezes", stands in a fine open position well raised above the lake and takes its name from the cool breezes that blow all the day under its trees.[70] Even though there are no streams, fountains or flower beds, its many *chinar* trees give it the great tranquil beauty.[71] Later during Shahjahan's reign the site of the garden was planted with hundred of *chinar* trees, on a regular grid. *Naseem bagh* is deeply impressive the great blocks of trees, foursquare almost as if hewn from masonry, is duplicated in the still water of lake, an immobile piece of geometry set in a matrix of ever changing clouds and water. Nasim, is a place for living in, and as the country took hold upon the imagination and affection of succeeding generations.[72]

To encourage the plantation of gardens in the empire Akbar also issued a farman to the Makhadims of the empire to plant gardens on 3rd July, 1578 A.D. According to this farman each makhadim had to build a mosque, a house, a chaupal, a garden, etc. in the village in the pargana assigned to him[73] Akbar was a great king who in benevolence was like Asoka and in his interest in art and gardening surely transcended him.

Emperor Jahangir (1605 – 1627) is the central figure in the garden art of India. Jahangir's Memoirs trace his developing skill, not only as a garden maker but also a plantsman and in the understanding of nature as a whole. He built splendid structures and gardens, the world's most famous and romanticised buildings. Jahangir's love of natural beauty was genuine and his aesthetic sense sometimes widened and almost expanded into a spirit of scientific enquiry, which was however, cramped by the empirical doctrines of his time and country. He made an arduous journey through the mountain passes to Kashmir to enjoy a view of the spring flowers there and during repeated visits recorded the names of the animals, birds and flowers he saw, distinguishing those which were not found in the plains of India, while the memoirs of Jahangir indicate that Jahangir left others to plan the buildings he required and shows that he took great delight in the arrangement of gardens in Kashmir and elsewhere.[74]

We know from the diaries of Sir Thomas Roe, ambassador of king James I (1603-1625) of England, who spent four years in India, that Jahangir exhibited strong interest in European portraiture, this interest combined with a love of the landscape of Kashmir produced in Jahangir a garden designer of no mean talent and, indeed the central proponent of Mughal garden art. Jahangir in turn was followed by his son Khurram, later to become Shahjahan, together they built the finest Kashmiri garden, the *shalamar bagh*. Nurjahan, his empress shared this design interest with Jahangir. Together they visited Kashmir many times, spending whole summers there, and the gardens some of which Jahangir started when still a prince, were progressively improved and enlarged by them. The royal gardens of *Shalamar*, *Achabal*, and *Verinag* belong to this period, together with Nurjahan's own garden on Lake Manasbal, known as *Jarogha bagh*. In his diaries, Jahangir describes the minutiae of the gardens, with details of flowers growing there and the general feasting and jollities of there summer days there.[75]

In the year 1607–8 Jahangir visited Kabul, and his visits to the gardens at Kabul are mentioned in his memoirs. there,

on this day I toured seven of Kabul's famous gardens ... First of all I walked round the Shahr-ara (city adoring), then the Mahtab (moon light) garden, then the garden that Bika begam, grandmother of my father had made, then through Urta bagh (middle garden), then a garden that

Maryam Makani, my own grand mother had made, then the Surat khana garden, which has a large chanar tree, the like of which there is not in the gardens of Kabul. Then having seen the char bagh which is the largest of the city gardens, I returned to my own abode... the Shahr-ara garden.. From time to time it has been added to and there not a garden like it for sweetness in Kabul.[76]

Near *Shahr-ara* garden Jahangir built a garden and named it as *Jahanara*. About its construction, he writes:

In the neighbourhood of this garden an excellent plot of land came to view, which I ordered to be bought from the owners. I ordered a stream that flows from the guzargah (ferry, also bleaching garden) to be diverted into the middle of the ground so that a garden might be made such that in beauty and sweetness there should not be in the inhabited world another like it. I gave it the name of Jahanara (world adoring).[77]

Kashmir was Jahangir's first love always. First time he visited along Kashmir with his father in 1589. He had scholarly instinct and love of nature. After he became the emperor of Hindustan, he visited the happy valley for a number of times for enjoying the cool and refreshing air of the valley. Both Jahangir and Nurjahan showed great interest in decorating the valley with a number of beautiful spots. Jahangir took many constructive steps to make the happy valley more beautiful by constructing world famous gardens. He also raise some structures at *Verinag* and *Achhabal*. He encouraged the plantation of *chinar* trees throughout the valley.[78] Jahangir has described Kashmir as a garden of eternal spring or an iron fort to a palace of kings – a delightful flower bed, and a heart expanding heritage for dervishes. He writes:

> *The garden-nymphs were brilliant,*
> *Their cheeks shone like lamps;*
> *There were fragrant buds on their stems (or under their rind),*
> *Like dark amulets on the arms of the beloved.*

The wakeful, ode – rehearsing nightingale whetted the desire of wine- drinkers;
At each fountain the duck dipped his beak like golden scissors cutting silk;
There were flowers - carpets and fresh rose buds,
The wind fanned the lamps of the roses,
The violet braided her locks,
The buds tied a knot in the heart.[79]

This beautiful description of Kashmir by Jahangir shows that how much he adored Kashmir. Across the lake from Srinagar lies most famous of Jahangir's gardens – the *Shalamar bagh*. This garden was built both by Jahangir and Shahjahan as a young man, the father using his flair for site and location, the son his exquisite taste for building.[80] Bernier who visited the garden in the reign of Aurangzeb describes this garden:

the most beautiful of all these gardens is one belonging to the king, called Chah-limar. The entrance from the lake is through a spacious canal, bordered with green turf, and running between two rows of poplars. Its length is about 500 paces, and it leads to a large summer house placed in the middle of the garden.[81]

Shalamar, most celebrated of the lake Dal gardens is said to date from the reign of Pravarasena ii (125-185 AD), who built a house here and called it *Shalamar* the "Abode of Love". But when the garden was built by Jahangir and Shahjahan the name was given as *Farah-Baksh*.[82] First laid out by Jahangir, and surely much influenced by Nurjahan, *Shalamar* combines a refinement of detail and proportion with an all pervading peace and calm. A Persian quotation appears at *Shalamar* which reads: if there be a Paradise on the face of earth, it is here it is here.

About the garden Jahangir writes:

Shalimar is also adjacent to the lake and has a beautiful water channel that comes from the mountains and empties into lake Dal. My son Khurram ordered it stopped up to create a waterfall one might enjoy. This spot is one of the scenic delights of Kashmir.[83]

The design of garden is small. At the head of garden, the river from the fields is diverted into a broad shallow canal leading to the wide rectangular basin in which the main black marble pavilion is set, surrounded on all sides by fountains. A great pavilion at the *zenana* garden provides the climax at *Shalamar* and seems like a magnet, to draw the whole of this garden to its heart – the central point from which four vistas open. After his accession, Shahjahan extended the design to the north and built a black marble pavilion in the *zenana* garden. The work was carried out in 1630 by Zafar Khan, the Mughal governor of Kashmir, and the new part of garden was named as *Faiz-Baksh*, the bestower of bounties.[84] It is far less showy than near by Nishat *bagh* and less worldly than Achhabal but justifies its reputation as one of the world's most beautiful garden.[85] This is another garden associated with Jahangir near Anantnag. Gushing out of a lime-stone mountain on which a forest of deodar is growing is a spring of clear water which is the life and soul of Achhabal garden. Supposedly the work of Nurjahan, it was originally called *Begumabad*. Smaller than both *Shalamar* and *Nishat*, it combines some of the qualities of both, being lively with strong rushing waters, as at *Nishat* but nevertheless achieving a feeling of serenity as at *Shalamar*.[86] Jahangir has given a description of this garden in his memoirs:

I camped on Tuesday the 31ˢᵗ [Sept.13] at the Achhival spring. This spring has ever more water than the others, and it has a beautiful waterfall. Around it are fine pine trees and elegant poplars whose top branches have grown together. Delightful places to sit have also been provided. As far as the eye could see was a splendid garden with ja'fari flowers in bloom. You'd say it was a patch of Paradise.[87]

The garden is laid out in terraces and is planted with fruit trees. A pavilion with rich cedar wood work presides over the canal which takes it birth in the mountain side spring.[88] Bernier was quite amazed at the beauty of this garden.

the garden is very handsome, laid out in regular walks and full of fruit trees – apple, pear, plum, apricot and cherry.[89]

Another garden was laid out by Jahangir at Vernag. Vernag (the powerful snake) about 11 miles from Achhabal in a direct line. A very lovely place, the garden was built by Jahangir in 1612-19, and it is said that it was designed and laid out by his wife Nur Mahal.[90] About its foundation Jahangir writes:

The source of Bihat is a spring in Kashmir called the Virnag; in the language of Indian a snake is Virnag…I went twice to the spring in my father's lifetime…It is an octagonal reservoir about 20 yards by 20 … the water is exceedingly pure … After my accession I ordered them to build the sides of the spring round with stone, and they made a garden round it with a canal; and built halls and houses about it, and made a place such that travellers over the world can point out few like it."[91] He further writes, "*what can be written of the purity of canal or of the greenery and the plants that sprout below the spring…In all of Kashmir there is no scenic spot so beautiful or charming as this one… It was ordered that plane trees should be planted on both sides of the canal.*[92]

Jahangir and Nurjahan reputedly loved Vernag above all others places. It is not hard to understand their reasons. It has a remoteness which makes the very act of arrival something to be treasured. Bernier visited this garden in 1665 and has generally admired it.

From Achivel I went yet a little more out of way to pass through another Royal garden, which is also very beautiful, and hath the same pleasantness with that of Achiavel, but this is peculiar in it that in one of its ponds there are fishes that come when they are called, and when you cast bread to them;

the biggest were of have golden rings in their noses, with inscriptions about them, which they say that renowned Nour-Mehalle, the wife of Jehan-Guire, the grandfather of Aureng-Zebe, caused to be fastened in them.[93]

Vernag is known for two surprises; at one end the casual entrance through the low arch, dark, unimportant and then the sunlit tank with its complex of domes and niches; its colour and stillness; at the bottom of the garden.[94] Jahangir loved this garden so much that he wished to be buried there. In 1636 Shahjahan visited this garden but he did not like the lay out and buildings of the garden. Malik Haider was appointed as *Daroga Bayutat* and directed to remodel the whole plan. The existing main channel divided the garden but two smaller channels were then built for irrigation of the side garden. A palace with a turkish bath was built in the centre. The garden was named *Shahabad*.[95]

Among the many mountain gardens of Kashmir is one at Manasbal lake called *darogha bagh* or *Lallarookh's bagh* containing a palace that some say, was originally built for Nurjahan. Set on terraced walls and planted with poplar trees, it just out into the clear water *"like some great high decked galleon"* as a hedge against the ravaging floods that often course through the region.[96]

On the western arm of Dal lake at Sadur Khan a garden known as *Bagh-i-baharara* said to have been laid by Nurjahan in 1623. This *bagh* has two terraces, one approaching the lake and other on a higher level, both with excellent views of water. Each of these terraces was prepared in the famous *Chahar Chinar* pattern with four *chinar* trees distributed evenly over a square plot in order to provide shade the whole day. A stone pavilion was built in the centre and water to irrigate the plants on all levels came from the Sind canals[97] Another garden Noor *bagh* was planted by Nurjahan in the vicinity of I'dgah. A branch of Shahkul canal was brought through Zunimar for the irrigation of the garden.[98] A garden was laid at Bijbihara by Jahangir's son Parvez. About this garden Jahangir writes:

This village has been bestowed upon my fortunate son Parvez. His agents have made a little garden overlooking the river and a small pavilion.[99]

The rulers were not alone in their love for gardens. The ladies of the court were also the enthusiastic creators of the gardens, and the nobility also took part in laying out them. Nurjahan took great interest in constructing buildings and gardens. According to Dutch traveller Pelsaert:

Meanwhile she erects very expansive buildings in all directions – serais, or halting places for travellers and merchants, and pleasure gardens and palaces such as no one has ever made before intending thereby to establish an enduring reputation.[100]

Besides Kashmir, she laid out gardens in many other parts of empire. One of the famous garden laid out by Nurjahan is the *Nurmanzil* garden. This garden has been praised by Jahangir:

As the praise of the garden of Nurmanzil and buildings that have been newly erected there continually reached me, on mounted my steed, and went to the stage of Bustan saray, and passed Tuesday in pleasure and at ease in that enchrancing rose garden. On the eve of Wednesday the garden of Nurmanzil was adored by the alig-thing of the hosts of prosperity.[101]

In late 1620 the *vakils* of Nurjahan completed a large sarai in Jalandhar district 25 miles east south-east of Sultanpur, 16 miles south of Jalandhar and a beautiful garden was also laid down there.[102] Jahangir makes a mention of this beautiful garden

on the 21st [Jan 2, 1621] camp was made in Nur saray. Here Nurjahan begam's agents had built a fine palace and regal garden.[103]

Two European travellers have attributed another garden round Agra to Nurjahan based on information they were privy to their journeys. In the latter half of Jahangir's reign, the Dutch merchant

Pelsaert noted that among many fort like gardens on the eastern bank of Yamuna *"handsome walls and great gateways"* were two that belonged to Jahangir: *"one named charbagh, the other Motimahal."* Although Pelsaert said nothing more about it, this *Motimahal* must certainly be *Motibagh* ("Pearl garden) mentioned by English traveler Peter Mundy, who in late 1632 noted the many gardens he saw in and round Agra. Of these he found, there were 3 that were

Darree ca bagh (Dehra bagh) and king Ecabers (Akbar's) on this side of river and Mootee ca bagh on the other side, the latter built by Noore Mohol.[104]

Nurjahan is also credited with the construction of the garden tomb of I'tmad-ud-daula at Agra. This tomb is a classic small scale example of old Persian plan. This Nurjahan's memorial to her parents stand on the river bank at Agra. The main gateway is approached by a straight drive, with orchards on either side, regularly planted.[105] The garden itself is in traditional *charbagh* form, with water channels dividing the square in four equal quadrants and the mausoleum placed in the centre. Nurjahan also got constructed garden tomb of Jahangir at Shahdara, Lahore. To the north-west of Lahore lies Nurjahan's old pleasure garden, once known as the *Dilkusha* garden. This place was visited by Jahangir during his life time, and after his death, he was buried there. Nurjahan designed his tomb, taking as her model the tomb of I'timad-ud-Daula, her parents' burial place at Agra.[106]

Nobility also took active part in laying out the gardens. Man Singh is credited with creating the Wah bagh at Hasan Abdal near Rawalpindi on the old route from lahore to Srinagar. Its romantic setting is said to have provided the Irish writer Thomas Moore with the inspiration for his famous poem *Lalla Rookh* (1817).[107] This garden palace was used as an imperial camping ground on the way to Kashmir.[108] Numerous other gardens were also erected at different places in the Mughal empire. After the death of prince Khusaru, Jahangir had his body disinterred, and brought down to Allahabad.

On the way, each resting place was marked by a shrine and a little garden was laid around it. He was buried in the Khusaru *bagh*, a terraced garden tomb at Allahabad.[109] Jahangir also got laid down a garden at the shrine of Khwaja Mui'nu-d-din Sanjari chishti.[110] A garden was also laid down by Jahangir at Nakodar. About its construction Jahangir writes:

I ordered Mu'zzu-l-mulk, the jagirdar of Nakodar, to erect a building and prepare a garden on one side of this embankment, so that wayfarers seeing it might be pleased.[111] Dutch merchant Pelsaert refers to the laying of the beautiful gardens at Lahore.

For some years however the present king has spent 5 or 6 of the cool months of each year in Lahore, and the city has now recovered, but more in splendour, royal buildings, palaces and gardens than in point of wealth.[112]

Pelsaert also praises two gardens in the town of Sikandara, which he writes belonged to the king (Jahangir).

There are also two gardens belonging to the king, one named Charbagh, the other Moti mahal and many more, with handsome walls and great gateways, more like forts than gardens, so that the city is most pleasantly adorned. Here the great lords far surpass hours in magnificence, for their gardens serve for their enjoyment while they were alive, and after death for their tombs, which during their life time they build with great magnificence in the middle of the garden.[113]

English traveller William Finch who visited India in 1611 has given a description of King's garden at Sirhind. According to him at Sirhind there was a fair tank, with a bridge of 15 stones, to connect the summer house built in the middle of it. From it a small canal was cut for the garden of the Emperor, which was situated at some distance. The way leading to royal garden was planted on both sides with trees and the garden itself was enclosed with a brick wall. It was planted with all sorts of fruit trees and flowers and was rented yearly for Rs.50,000. It was divided into 4 squares, each a *kos* in length.[114]

Jahangir's mother also laid out a garden at the *pargana* of Jusat, in 1613 which Jahangir inspected a few years later.[115] During the reign of Jahangir construction of gardens was intensified on a large scale and the ladies of Mughal harem as well as noble took active part in plantation of gardens.

Shahjahan's [1628 – 58] reign is considered as the Golden Age of Mughal art and architecture. Among the buildings and gardens he erected were the unsurpassed Taj Mahal, the Pearl Mosque in Agra, *Shalamar* garden Lahore and many more. It is known that Shahjahan showed keen interest in the construction of gardens in Kashmir right from his father's reign. Shahjahan's great genius was for building and planning, though the gardens in which edifices were sat were no less important than those of his father in Kashmir. However, being erected on flat ground and encompassing the structure, they returned to the Persian *Char-bagh* tradition, although they were now much enlarged and the narrow water channels became smooth canals.[116] Shahjahan had so much passion for gardens that he spent about 25 million rupees in the construction of grand buildings such as Masjids, palaces, forts, tombs, hunting retreats, and gardens in Delhi, Agra, Lahore, Kabul, and others parts of empire.[117]

One of the most important project of Shahjahan in terms of the erection of gardens was the laying of the gardens of *Farah Baksh* and *Faiz Baksh* (Shalamar bagh) at Kashmir. Inayat Khan writes in his *The Shahjahan-nama*:

Towards the close of the 14th regnal year (Feb.1620), His Majesty (Shahjahan) proceeded to paradise the like Kashmir in order to accompany the emperor (Jahangir) and visit the places of excursion… In these days, the Shalamar garden, which is famed throughout Kashmir for its kausar – like canal, its terraces and its buildings, were planned and completed under the supervision of His Majesty.[118] He further writes "..*the bagh-i-Farah Bakhsh constructed by His Majesty's command during the Late emperor Jahangir's reign, … the garden is irrigated by a canal resembling the kausar, the nectar-flowing river of paradise. It is 10 yards wide, but as it circulates among the beds, it is contracted to a breadth of 3. After flowing through that villa in the middle of the garden, and after supplying countless reservoirs and fountains, it expands as it glides through the picturesque domains for another 1,000 yards. It passes lovely avenues of plane and poplar trees gracing both banks, until its waters are eventually merged in the Dal. Thus a boat from the lake, by passing along this canal and traversing the above distance, enters at once within the limits of the grounds. In these days by the sublime command, at the rear of the Farah Baksh garden, another garden was laid out which was designated Bagh-i-Faiz Bakhsh, and it was directed that the terraces should be constructed so as to admit of the great canal flowing through at its full breadth.*"[119]

Inayat Khan writes that another garden was laid out by Shahjahan at Shahabad (Vernag).

On the 5th of Rabi' 11, 1044 (28th Sept., 1634), the imperial camp moved from Sahibabad to Vernag, which is the last of the natural springs, and a total of 18 imperial kos distant from the city. This spring is the source of river Behat, and is famous for intense coldness… at the back stands a hill of the most picturesque form and appearance, clothed with luxuriant vegetation and stately forest trees overshadowing the rippling brooks. Here His Majesty constructed a delightful garden provided with summer houses, baths, reservoirs, cascades, and fountains, and he named the place Shahabad[120]

Another garden was laid out in Kashmir on the instructions of Shahjahan - *Chashma-i-Shahi*. It is the smallest of the Mughal gardens and is the closet to town (8 Km from Srinagar). Its name means "the royal spring" and the ice cold water from the spring is reputedly filled with excellent medicinal properties and is beautifully located with a view of Lake Dal and the surrounding hills.[121] Built in 1632, an early inscription at the gateway attributes this garden to Shahjahan himself. It seems more likely that the actual builder was 'Ali Mardan Khan, working on Emperor's instructions.[122] Ali

Mardan Khan, a Persian, had great engineering skill. And it was he who designed and built this small but elegant garden of *Chashma-i-Shahi* overlooking the Dal lake.[123] *Chashma-i-Shahi* is a famous spring of ice cold water. The architecture of the garden pavilion is in harmony with the pyramided peaks of the rugged mountains which provide the background. Its roofs repeats the conical beauty of the mountains in a striking manner. The pavilion enshrines the spring whose water is relished for its purity and digestive qualities.[124] the Mughal work can be seen in the cascades, the plinths of the pavilions of the garden and also in the complex of canals, tanks and fountains.[125] Dara the eldest son of Shahjahan inherited his father's eye for splendour and his grand father's love of nature, as his choice for site above Dal lake – on which he got built a school of astrology for his tutor, Akhund Mullah shah, was inspired. High up the mountainside behind *Chashma-i-Shahi* is *Pari Mahal*, the 'Fairies Palace', and it lived up to its magical name in its siting. Originally a rather grandiose layout, with a central building and flanking pavilions, it sits upon a podium with, behind, stepped terraces rising to a central pavilion[126] *Pari Mahal*, is on the hill immediately south-east of the Dal lake. There is a suggestion of Greek temple in its siting, for it lies across a spur of rock, thrown in to relief against the darker, higher mountain behind[127]. The terraces are made in brick masonary.[128] *Pari Mahal* has the special magic of inaccessibility. The foreground is dramatic: white-stemmed poplars rise from the level lake shore, and behind and above them the Fairies Palace seems to appear and disappear as the light strikes or leaves the wall.[129] One more garden was laid down by Dara Shikoh some 29 miles from Srinagar to the south in the town of Bijbehara. Here on both sides of the river Jehlum was the garden of Dara Shikoh, in which there were magnificent *chinars*.[130] The two portions of the garden were united by a stone bridge. Present name of the place is now Wazir *bagh*.[131] Traces have been found of the parterres and canals, but the garden fell in to ruin, and is today marked only by the great *chinars*, and by the remains of a pavilion and a bridge.[132] The court nobility also vied with the garden loving rulers in Kashmir: the Zafarbad garden created by Zafar khan is an especially interesting result of this rivalry. Zafar Khan, an excellent governor of Kashmir, who was married to one of Mumtaz Mahal's nieces, wrote about his gardens in a Persian *mathnawi*. A few of the miniature illustrations to the text record the fact that Shahjahan once paid a visit to this garden.[132a]

In 1638 Shahjahan moved his capital from Agra to Delhi (Shahjahanabad).[133] In Delhi, Shahjahan founded a new city called Shahjahanabad.[134] Named after its builder, it was perhaps the finest succession of cities which have been built at Delhi. Bernier, describing Shahjahan's decision to develop a new city at Delhi, says that the emperor considered that the summer heat and humidity of Agra, made it unsuitable for his capital.[135] Numerous gardens were laid out in the new city. About the city of Shahjahanabad Sujan Rai Bhandari writes in his Khulasat-ut-Tawarikh:

"In the year A.H. 1048, being the 12th reigning year of the king, Emperor Shahjahan built, a city near the old City of Delhi and named it Shahjahanabad. Its fort was built entirely by red stone. It has lofty buildings, wide canals, fountains, gardens, flower beds and in all the year round full of green trees, flowers and fruits. It has a wide moat that is full of water all the time... In the East of it the river Jamuna washes the walls. A canal from the river was separated and taken through the bazars, land and by-lanes to pass through gardens and orchards."[136]

In the year 1639 Shahjahan started the construction of the Red Fort. It took some 9 years to complete it.[137] The whole fort has been carefully designed in every detail, making adequate and suitable provision for the requirement of all the functions. There were separate halls for public and private *darbars*, a complex of palaces surrounded by well laid out gardens for the Emperor and the royal family.[138] Among the well defined gardens, each with its own characters were the *Hayat Baksh* or life giving garden, and the *Mahtab Bagh* or Moon light garden, combined to form one grand design. The latter no longer exists, but much of the *Hayat Baksh* remains. Designed as a water garden, the central pavilion stood in a pool full of fountains.

From this ran four canals, which terminated at south and north in small water pavilions, the *Sawan* and *Bhadon*. These names are linked with the months of July and August, when the ladies sat in the pavilion to enjoy the prevailing breeze.[139] About the plantation of the gardens in the city of Shahjahanabad, Niccola Manucci writes:

"Shahjahan planted two large gardens, one on the north side, the other on the south side, and for the reason that river Jamnah does not rise enough to permit of its irrigating these gardens, Shahjahan at great expense and labour, constructed a deep canal from a river adjacent to the city of Serend (Sirhind), one hundred leagues from Delhi."[140]

Jahanara *Begum*, the favourite daughter of Shahjahan on whom the emperor had conferred the title of *Begum Sahib*, took keen interest in the development of Shahjahanabad. In 1650 she laid a lovely and spacious garden in the heart of the city which not only was a source of continual pleasure but also served the utilitarian purpose of green lung. She did everything she could to make *Begum ka bagh* a thing of beauty – with pools, channels for running water, lovely fountains with faucets variously fashioned of silver; canopies (*chhatris*). There were many flowering and fruit trees in the garden. Ali Mardan Khan's canal flowed through it, providing plentiful irrigation on round the year. When Delhi was occupied by The English East India Company, The name *Begum Ka Bagh* was changed to company *bagh*; and after 1857 when queen Victoria was proclaimed Empress of India, its name was changed to Queen's garden. What remains of it today is the Gandhi ground.[141] Another garden was laid by Begum Sahib's sister Roshanara, the Roshanara *bagh*. Roshanara, favourite sister of Aurangzeb, lies buried in her own garden house an elaborate white pavilion with creeper clad walls standing on a low wide platform in the centre of the upper terrace in the garden, still called by her name.[142] A garden was also laid down by Sarhindi *Begum* at Shahjahanabad. It is adjacent to Roshanara *bagh* at Delhi.[143]

Among many gardens laid down by Shahjahan was the Tees Hazari *Bagh*, outside Kashmere gate, in the traditional style, with large *neem* trees all round the periphery and other auspicious trees on each point of compass. Ali Mardan's canal, before entering the Kabuli gate, ran through this garden, abundantly irrigating it. This spacious garden later became the *jagir* of Jahanara and was well tended by her. Later on this garden was given to Aurangzeb's daughter Zeb-un-nisa as *jagir*. She grew to be very fond of this *jagir* and was ultimately buried there after her death. Malika-i-Zamani Begum, the daughter of Farruksiyar, (Aurangzeb's grandson) who had been married to Mohammed Shah Rangila, also took a fancy to this garden and was laid to rest there.[144] The most outstanding work of one of Shahjahan's favourite wives A'Azuu-un-Nisa by which she is remembered is the creation of *Shalamar bagh*, Delhi, a garden with some lofty and beautiful buildings, some six miles from Delhi on the road to Karnal. This garden was modelled on the pattern of Emperor's gardens of the same name in Kashmir and in Lahore.[145] Its remains are recorded in a plan showing a garden of considerable size, its main feature being a central canal about 18 feet wide, which ran the full length of the garden. According to a contemporary historian, Muhammad Saleh, it was originally even larger. The upper terrace stood some nine feet, above the lower, and at the change of level was a complex of tanks and buildings, with rows of pearl showering fountains and marvellously adorned halls. Some distance below lay an octagonal reservoir, similar to that at Vernag, with a third row and even larger terrace below it.[146] About this garden Inayat Khan writes:

… in conformity to the ever obeyed farman, a royal garden had been founded outside Shahjahanabad at a distance of two and one – half ordinary kos from the precincts of the palace, on a plot of ground in a locality called Badli. This garden had been bestowed by His Majesty as a gift to his consort Akbarbadi Mahal and is now designated A'izzabad. In these days after a lapse of 4 years and an outlay of 10 lakhs of rupees, the garden was now completed together with its serai, which had been erected of sundried bricks, but was subsequently rebuilt with baked bricks and morter at the lady's own expense.[147]

There was a fine mango grove shading the highest pool, and a picturesque lake over grown with lotus, which suddenly led to the chief building, the palace called *Sheesh* Mahal. It was in this *Sheesh Mahal* that Aurangzeb had crowned himself emperor.[148] It took four years to build this garden and was opened ceremonially by Shahjahan in September 1650, the occasion was being marked by a festival. Bernier who visited this garden describes it as regal and beautiful.[149] The *Shalamar bagh* was in fact used as a holiday resort; when once Aurangzeb set out in a long journey, lasting over a year, he spent six days here while the preparations being completed. After his invasion of India, Nadir Shah too celebrated a holiday in the *Shalamar bagh*, before leaving Delhi. When British occupied Delhi in 1803, the *Shalamar bagh* was retained as a summer retreat and weekend holiday spot.[150] Chandra Bhan Brahman, a contemporary historian of Shahjahan has greatly admired the gardens in the city of Shahjahanabad:

beautiful buildings, domes and gardens that are green and refreshing.[151]

From the account of Chandra Bhan Brahman one finds that numerous gardens were laid down in the city of Lahore also.

The city is well decorated, flowers and fruits there are in abundance... There are a large number of gardens in the vicinity of the city like Bagh-i-Dilkusha and Dil Amez, belonging to Begum Sahiba-kalan, Bagh of Mirza Kamran and Naulakh bagh etc. However Bagh-i-Faiz Bakhsh, constructed during the reign of Hazart Sahib-i-Qiran-i-Sani (Akbar) is unique among all gardens...It is custom that during spring season every year, when the flowers are in full bloom, the emperor visits the garden.[152]

And in the plains of India the most notable example of a Mughal garden is the *Shalamar bagh* near Lahore, built by Shahjahan in 1637. It is designed on the same principles as those governing the plans of most of these pleasances.[153] About its construction Inayat Khan writes:

...His Majesty issued an order directing that the royal architects and other able and engineers in this field should choose a site for planting a garden in the vicinity of Lahore, on the banks of the canal leading to Lahore. The proposed garden was to contain both high and low ground and be capable of having a number of reservoirs, canals, fountains and cascades constructed within it on different terraces. Accordingly, a spot was selected which was so delightfully adapted to the purpose that it was universally commended. And in a lucky moment, on 3rd of Rabi 1105 (12 June, 1641), skilful artificers commenced its construction.[154]

This garden was laid by Ali Mardan Khan on Shahjahan's instructions. The first great undertaking had been the construction of a canal to bring the waters of the Ravi river up to the gardens in Lahore. By 1633 the canal was ready and Shahjahan celebrated its completion by giving instructions for the creation of a garden on a grand scale, with tanks and fountains, a bath house and several pavilions. About 1642 the work was complete and Shahjahan paid a ceremonial visit to this garden.[155] About its completion Inayat Khan writes:

As the architectural embellishments of the superb and delightful gardens Bagh-i-Faiz Bakhsh and Bagh-i-Farah Bakhsh had been completed about this time, in a chosen moment during the early part of Sha'ban 1052 (31 Oct.,1642), His Majesty made a pleasure excursion to those paradise like terraces. And the garden and the agreeable pavilions which had been erected about the grounds, which all vied with the heavens in grandeur, were now found suitable to the royal taste. In fact never before had a garden of such a magnificent description had seen or heard of; for the building alone of this earthly paradise had been erected at an outlay of 6 lakhs of rupees... Ultimately, at the suggestion of several learned specialists who possessed great engineering skill, use was made of only 5 kos of the canal laid out by Ali Mardan Khan's men; and a new channel 32 kos long was excavated by which a plentiful supply of water reached the garden without any Impediment.[156]

This garden is formed by means of a series of rectangular terraces arranged in descending levels with the object of maintaining a continuous flow of water throughout the entire system, as fountains, pools, basins, cascades and similar devices are so distributed among the parterres as to make the whole in to a very effective type of water garden.[157] The upper terrace of the garden was known as the *Farah Bakhsh* (the bestower of pleasure), while the middle and lower terraces, forming the more public part of the garden, were known together as the *Faiz Bakhsh* (the bestower of plenty).[158] *Shalamar* is a garden of great beauty characterised by balance and symmetry. Praising beauty the beauty of this garden Sujan Rai writes:

Although there are many charming gardens and a thousand pleasant roses bowers in the outskirts of the city, yet the garden of Shalimar, which the emperor Shahjahan laid out in imitation of the garden of the Kashmir, ravishes the heart of the beholder[159]

A number of gardens were also laid down in the palace-fortresses constructed by the Mughals. The fort at Lahore is one of three great Mughal palace forts. Here Akbar, Jahangir, Shahjahan and Aurangzeb made their contributions, but it is much changed by time. Two gardens survive here in recognisable form. Jahangir's Quadrangle, and the *Paien bagh* or *zenana* garden. Jahangir's quadrangle is a large open space with a central reservoir containing fountains and a marble platform reached by a causeway. The *Paien bagh* is more intimate in scale, with paths paved in a hexagonal pattern of red bricks. Here the ladies of the harem could walk daily and enjoy the fruits and fragrance which were the theme of the garden.[160] The two principal gardens were laid down in the Agra Fort, the *Machchi Bhavan*, or fish square; and the *Anguri bagh*, or grape Garden. The *Machchi Bhavan* owes its name to the pools of sacred fish which it once contained. The other garden *Anguri bagh* was the private garden of the *zenana* quarter, it lies open and level and sunny; a traditional Mughal garden.[161] About the gardens in the Mughal palace forts,

Niccolao Manucci – an Italian gunner and part time doctor in the service of Dara Shikoh – writes that they were

full of gardens with running water, which flows in channels into reservoirs of stone, jasper and marble. In all the rooms and halls of these palaces there are ordinarily fountains or reservoirs of the same stone of proportionate size. In the gardens of these palaces there are always flowers according to the season. There are no large fruit trees of any sort, in order not to hinder the delight of the open view[162]

A garden was also laid down at Kabul during the reign of Shahjahan. Inayat Khan writes:

On the 18th of Sha'ban 1056 (29 Sept.1646), the Bagh-i-Safa (at Kabul) acquired new freshness by the auspicious arrival. A sublime farman was issued for the planting of a new garden at locality Nimla, which is the most charming spot, and for the construction of a canal 4 yards wide, by which the stream near the town could be made to flow through the grounds and summer houses"[163] He further writes *"On the 10th of sha'ban 1057 (10th Sept.1647), the recently completed garden at locality Nimla, the planting of which had been commenced during the past year, was graced with the August presence and received the name of Bagh-i-Farah Afza.*[164]

One of the most beautiful garden tomb erected in the reign of Shahjahan is the Taj Mahal. All architectural experiences, beautiful though some of the results undoubtedly were, recede into the background when compared with the materialised vision of loveliness known as Taj Mahal, a monument which marks the "perfect moment" in the evolution of architecture during the Mughal period. This building which stands on a bend in the river Jumna at Agra, is the mausoleum of the emperor Shahjahan's well beloved consort the empress Arjuand Banu *Begum*, whose titles Mumtaj Mahall (chosen of the palace) and Taj Mahal (crown of the palace) have been abbreviated into the Taj. Although the tomb building itself was the raison d'être of the undertaking, the main structure actually occupies only a relatively small portion of the architectural scheme as a whole. The plan of the whole conception takes the form of a rectangle aligned north and south and measuring 1900 feet by 1000 feet with the central divided off into a square garden of 1000 feet side. The entire garden portion, including the tomb terrace, is enclosed within a high boundary wall having broad octagonal pavilions at each corner, and a monumental entrance gateway in the centre of the southern side.[165] The garden is a *char-bagh*, a four fold field plot, with a tank of white marble in the centre.[166] The building or the tomb is placed not at the centre of the garden but at the end of the garden on a raised terrace to form the climax of the whole design.[167] But the ornamental gardens were so planned as to prepare the spectator for the exquisite appearance and lovely dignity of the central structure, in addition there were water courses with fountains and elevated lotus pool, all arranged to mirror the beauties of Taj from various points of view.[168] According to Bernier:

Resuming the walk along the main terrace, you see before you at a distance a large dome, in which is the sepulchre, and to the right and left of that dome on a lower surface you observe several garden walks covered with trees and many parterres full of flowers.[169]

Inayat Khan writes that a vast amount of money was spent on the maintenance of Taj Mahal

To maintain the mausoleum and its garden, His Majesty established an endowment consisting of the annual revenues of 30 Hamlets situated in the pargana of Akbarabad and a few others, which amount to 40 lakhs of dams, equivalent to 1 lakh of rupees, or more during favourable years[170]

From the account of foreign travellers on finds that beautiful gardens were laid down in almost all the parts of Mughal Empire. About the garden in the city of Sirhind, De Laet writes:

The City[171] *has a beautiful tank in the middle of which is a temple approached by a stone bridge of 15 arches: a cos distant from this tank is a royal park approached by a canal and a paved road, 50 feet broad, between the fine avenue of trees. This park is square shaped, each of its sides being more than a cos long. It is surrounded by a brick wall, within all kinds of fruit trees, plants and flowers. Some writers say that 50,000 rupees are yearly spent on this park. In the middle were broad pathways bordered with tall Cypress trees meet in the form of a cross, a royal palace is to be seen, beautifully built, and surrounded by fine colonnade.*[172]

A garden was constructed at Ujjain by Aurangzeb and Murad Bakhsh as a mark of their victory over Maharaja Jaswant Singh Bahadur during their war of succession. They entered the city of Ujjain with their victorious army and they transformed it into heaven and named it as *Dar-ul-Fateh* and they issued the order that the palace should be decorated with an inn (Serai), a garden and should be called by the name of Fatehabad.[173] Mughal emperor Aurangzeb's (1658–1707) contribution to architecture was however, considerable. It consisted largely of mosques, and of these, the Moti Masjid in the Red Fort at Delhi is of outstanding quality. Of landscape architecture it is not unfair to say that he was unaware of its existence as an art and the few gardens laid out in his reign show clearly how the standards of design had begun to degenerate. Yet the *Shalamar* Bagh at Delhi must have held attraction for him, for it was here that he was provisionally crowned as emperor, upon the deposition of his father.[174] It is mentioned in *Massir-i-Alamgiri*:

As astrologers indicated Wednesday the 21ˢᵗ July/ 1ˢᵗ zilq as an auspicious day for accession, and no time was left to go to Delhi and make preparations for the ceremony, Aurangzeb stayed for a few hours in the garden of Aghrabab [Shalamar], and sat on the throne at the above mentioned auspicious time. The presents which were made to the princes, the high grandees, the mansabdars and other officers, are beyond numbering.[175]

In the context of garden making, Aurangzeb's long reign is of interest chiefly for the light it throws on the achievements and customs of his predecessors. The zenith had been reached with Shahjahan, Aurangzeb's own contribution was minimal.[176] However, the accounts of the Travellers like Bernier, Tavernier, and Manucci are of great importance as they provide the details of the gardens in the Mughal empire. Bernier especially left the most complete description of the gardens of Kashmir.[177] With the decline in royal patronage, the hobby of gardening became a pastime for some of the princes, nobles and generals. That they loved their gardens is evident from some of the portraits which are provided with a background of flowering peaches. One also comes across numerous paintings in which princes are shown in their gardens.[178] In Deccan Aurangzeb set up a city called Aurangabad as his southern capital, convenient for his military campaigns. Here he built a citadel, the *kila Arh*, and two gardens were built here. One, the *Pan Chakki* or water-mill, contains the tomb of a *Chishti* saint who was his spiritual mentor, while in the north-east of the town lies the tomb garden of his wife.[179] According to Tavernier:

Formerly this town had another name, and it is the place where Aurangzeb, who reigns at present, gave battle to his brother Sultan Shuja, who held the government of whole of Bengal. Aurangzeb being victorious gave his name to the town, and he built there a handsome house with a garden and a small mosque.[180]

Manucci given a vivid description of the garden in the laid down in the city of Lahore during the reign of Aurangzeb.

Round the city are fine gardens filled with various kinds of fruits.[181]

At Lahore Aurangzeb's Badshahi Mosque had a garden. His daughter Zebunissa Begam laid down two gardens here, the Chau-Burgi bagh and the Nawan Kot bagh, but little trace of either remains.[182] However the period from 1707 to 1750 was full of turmoils and constant warfare. During this period patronage declined, art, architecture and gardens also deteriorated. Bahadur Shah Jafar (AD1707–1712), the successor of Aurangzeb, extended patronage to artists and also loved gardens and nature in all her verifying moods. He laid out two gardens, one below the palace wall at Shahjahanabad and one in Shahdara.[183] During his reign, the festival of *Poohl walon ki sair* (festival of flower sellers) reached its peak in the participation of court and the people of all the communities. The king would himself go to the '*jharna*' – (a beautiful garden with – cascades and fountains, originally built by Firoz Tughlaq (1351–1388) to celebrate the festival.[184] During the reign of Jahandar Shah (1712–1713) confusion and disorder prevailed. While his predecessors planted trees and raised gardens, he caused their wide spread destruction.

An order was given to cut down all the lofty trees from the palace to the hunting preserve called Jahan-Numa. Khush hal Chand, a rare instance of an Indian taking notice of the beauty of natural objects, laments over the wanton destruction of the spreading trees, with heads reaching the sky, the refuge and solace of the weary, foot-sore, traveler, the abode of far-flying and sweet singing birds. Throughout Delhi and its environs it was for the trees like the coming of judgement day: and the trees on the two banks of the Faiz canal, planted by emperors of the high empire, ceasing to raise their heads to Heaven, received wounds in the garment of their existence and fell in to the dust of degradation and disgrace.[185]

Farruksiyar *(1713–1719)* was another unworthy ruler and we find no record of his interest in gardening.[186] But there was some sort of passion for gardens in later Mughals for instance there are certain references which show that Ahmad Shah (1751– 52) used to go for walk in *Anguri Bagh* and garden of Ali Mardan Khan.[187]

Right from the beginning of the Mughal rule, apart from the rulers, Mughal nobles and the Provincial rulers also also made a significant contribution in the plantation of gardens. When Babur began to build gardens, he encouraged his chief nobility to imitate his example and the amirs were not slow in carrying out the wishes of their royal master. His followers profited by his example:

Khalifa also and Shaikh Zain, Yunas-i-Ali and whoever got land on that other bank river laid out orderly and regular gardens with tanks, made running water also by setting up wheels like those in Dipalpur and Lahore. The people of Hind who had never seen grounds planned so symmetrically and thus laid out, called the side of the Jun, where our residences were, Kabul.[188]

Mirza Haidar Dughlat, the contemporary of Humayun has explained that the land in Kashmir was divided into 4 kinds. The cultivation was: (I) by irrigation, (2) on the land not needing artificial irrigation, (3) garden, (4) level ground[189] which shows that gardening was regarded as important activity in Kashmir. Akbar also laid out so many gardens at other places. In the village of Kakrali, 7 miles to the south of Agra, being attracted by its beauty, he built a palace to serve as a hunting lodge and pleasure house. The courtiers were encouraged to build houses and gardens and a small town sprang up round the palace.[190] In *Tabqat-i-Akbari*, building of gardens and fort at village Mulathan by Akbar is mentioned. It is written:

the emperor on his return journey, when he reached the neighbourhood of Ambir [4 miles north east of Jaipur], ordered a fort and town to be founded in the Mulathan, one of the dependencies of Ambir ... walls and forts; and gates and gardens were allotted among the amirs, and injunctions were given for the completion of the work. So a building (imarat) wqhich might have taken years for its completion, was finished in 20 days.[191]

Abdu'r Rahim *Khan-i-Khana* laid down a garden at Burhanpur. William Finch has given an account of this garden, he has written that this garden was called *Lal Bagh* and was situated 2 *kos* from the city of Burhanpur. The whole way from town to garden was planted with shady trees. It had many ravishes, a small square tank between 4 trees, and a banqueting house. It was enclosed with a wall.[192]

Another garden made by *Khan-i-Khana* near Sarkhej is mentioned by Nicholas Withington (1612– 1616). He has written that about a mile and half from Sarkhej; there is situated a very fair garden, built by *Khan-i-Khana*, as a monument of his victory over the kings of Gujarat.[193] The remains of this *bagh* called *Fath bagh* or garden of victory, laid out in 1584 can still be seen near lake of Sarkhej.[194] Details of Fath bagh has also been given by Jahangir,

this garden is situated on the ground on which the commander-in-chief, Khana Khana Ataliq fought and defeated Nabu. On this account he called it Bagh-i-Fath; the people of Gujarat call it Fath bari.[195]

Nooruddin in *Latifa-i-Faiyazi* writes that Salabat Khan built a garden in the city of Ahmednagar[196] and it was named as *Farahbakhsh*.[197] This garden is also mentioned by Amin Ahmad Razi in his work *Haft Iqlim*, he has described this garden having exceptional beauty. He has also mentioned the gardens of Golkonda.[198] Father Monserrate writes that the houses of the nobles and rich people were adorned with the gardens with fruit and ornamental trees.

None the less the rich adorn the roofs and arched ceilings of their houses with carvings and paintings, plan ornamental gardens in their courtyards[199]

He has also praised the parks and gardens of Sirhind.[200] Mohammad Amin Qazvini has mentioned that princess and nobles along with other rich persons, have built beautiful and spacious houses and gardens in the city of Akabrabad on both sides of river Jamuna.[201] While staying at Lahore in the attendance on the emperor Akbar, Nizammu'd-din Ahmad laid out a garden and in this garden he was buried after his death.[202] In Dholpur Sadiq Khan, a Mughal noble built a serai and had a wide and beautiful garden attached to it[203] In Goa and Gulbarga [204] numerous gardens were planted by Provincial governors. Ralph Fitch has described Goa as principal town occupied by Portuguese and its viceroy has maintained the city very well. It was full of orchards and gardens.[205] There are certain inscriptions of Akbar's period which show that gardens and orchards were planted by the Nawab of Gulberga, Ibrahim Adil Shah, one of this is on a well in the fort of Gulberga dated 13th September 1585 on which it is written that Nawab Adil Shah made a pilgrimage to the shrine of Sayyid Muhammad Husaini Gesu Daraz, here at this place *Makhduma-i-Jahan*, mother of the said Nawab caused a well to be dug and a garden to be planted. Then there is another inscription in a well in Adilabad a suburb of Gulburga dated 13th December 1585 to 1st Dec. 1586 AD on which it is written that Zabit Khan, the lieutenant of the city of Ahsanabad, built a house and a garden in Adilabad.[206] All this shows that plantation of gardens was a passion not only of rulers but of nobles and provincial governors also. Mughal noble Muqarrab Khan laid down a garden at Kairana (in sarkar Saharanpur). This garden has been greatly praised by Jahangir.

It is very fine and enjoyable garden. Within a masonry, (Pukhta, pucca) wall, flowerbeds, have been laid out to the extent of the 140 bighas…There is no kind of the tree belonging to a warm or cold climate that is not to be found in it and cypresses of graceful form, such as I have never seen before. I ordered the cypresses to be counted, and they came to 300.[207]

Asaf Khan, the brother of Nurjahan had a great passion for gardens. He spent a fortune of 40 million rupees in erecting buildings and gardens. He was also responsible for constructing canals and h*aveli* garden. His *haveli* garden which was exceedingly beautiful and had beautiful scented flowers was visited by Jahangir, who greatly admired this garden:

As Asaf Khan had represented that his haveli garden was exceedingly green and pleasant and all sorts of flowers and scented plants have bloomed there, at his request I went to it . . . In truth it was a very nice villa and I was much pleased. [208]

Asaf khan also laid down a garden at Lahore. This garden was visited by William Finch. It was situated on the east side of the Lahore fort. It was

small, neat with walks, diverse tanks and jounters; as you enter, a faire devoncan supported with stone pillars with a faire tank in the midst . . . Beyond are the other galleries and walks, diverse buildings for his own women neatly contrived, and behind a small garden and garden house. [209]

A garden was also laid out by Asaf khan at Kashmir the *Nishat bagh,* a little way along Lake Dal from *Shalamar,*. Siting of the garden is far more spectacular than the *Shalamar,* being tucked more into the mountain, and it has a far steeper ascent, so contains large features as well.[210] The selection of the sight reflected the mature and artistic taste of Asaf Khan. Nishat has twelve terraces for which it was known as *"Garden of terraces"*[211] From the shore of lake its twelve terraces rise dramatically higher and higher to the mountain side. Each terrace represents a Zodiac sign. At the head of every waterfall is a stone or marble seat. *Nishat* provides a ground view of Lake and mountains.[212] Emperor Shahjahan visited this garden and was much delighted by its scenic beauty. Asaf Khan laid out another garden in *'Mahcchi Bhawan'* in Kashmir during the reign of Jahangir[213] Numerous other gardens were laid down in Kashmir under the patronage of Mughals. Some of the celebrated gardens of Mughals have been mentioned by Inyat Khan in his *The Shahjahan-nama*: *Bagh-i-NurAfzai, Bagh-i-Baharara, Bagh-i-Aishabad, Bagh-i-Jahanara, Bagh-i-Nur-Afshan,* commonly called *Nurbagh,* which is situated on the bank of the Behat (Jehlum), *Bagh-i-Shahabad, Bagh-i-Murad,* the 3 gardens *Bagh-i-Nasim, Bagh-i-Afzalabad,* and *Bagh-i-Saifabad,* which stand contiguous to each other on the margin of the lake Dal, and finally *Bagh-i-illahi,* which is watered by the canal Lajmakul that flows through the populous quarters of the city. In the midst of this garden, and by the side of this canal, stands a huge and ancient plane tree (*chinar*) whose truck even 8 stalwart men could not encircle in their embrace, as it measures 14 yards in circumference.[214] Another garden *Bagh-i-Husnabad* was built by Zafar Khan.[215] A number of fruit gardens and leisure gardens were built by Ali Mardan Khan at Sodhra.[216] Sujan Rai Bhandari writes:

Sodhra is an old fort on the bank of river Chenab. In the reign of Emperor Shah Jahan Ali Mardan Khan, the premier noble founded a city near the aforesaid villages and named it Ibrahimabad after his own son. He laid out a pleasure garden, which rivals the garden of Shalimar, and built lofty houses, spending 6 lac of rupees on these buildings, garden and a canal[217] *which brings the river Tavi to the garden-house. The imperial government has assigned to the premier noble 2,000 villages of Sodhra rent free for the repair of the aforesaid garden and city.*[218]

In these gardens of Ali Mardan Khan the breeze was laden with fragrances. The fruit gardens like *"Manglawala"* and *"Atharwala"* were built on the southern side of Sodhra, where as leisure garden of Ali Mardan Khan was on the eastern fringes. The gardens were enclosed by periphery walls with watch towers on each corner but their remains have now been completely obliterated. While praising the beauty of the garden Faaiz says in his famous book *"Kuliyal Faaiz"*:

... its entrance is broad like the forehead of a damsel and the beauty of which surpasses that of a garden itself... its vastness encompasses the whole earth, none could ever see the like garden... its fountains role the pearls and remain in the continuous flow... The flower laden garden swells exultation and pampers an heart like a cup of wine...[219]

One of the best surviving garden laid by a Mughal noble is at Pinjaur near Kalka. This garden was built in the reign of Aurangzeb by his General Fidai Khan in 1661. For his model he had the *Shalamar* garden at Lahore.[220] According to Sujan Rai:

In the 4th reigning year of Alamgir, Fidai Khan who was one of the Umara decided to make this place his watan. He called it Pinjaur. The Raja of this place had his riyasat. He was removed. There he constructed a terraced garden, containing beautiful buildings with a canal flowing through it. This canal was brought from the nearby mountains. Large fountains played in the garden. Roses of this garden are very famous. I visited this garden during spring. 40 monds of roses according to wazan-i-alamgiri were collected in the Gulabkhana. It was not the peak of the season.[221]

There are six terraces in this garden with a canal running through. At the end of the first terrace is a white building called *shish Mahal* which appears like a huge transparent shell against the dark blue evening sky. At the end of the second terrace is a two storied building called the *Rang Mahal*, or the painted palace, flanked by guest house. Pinjaur is truly the most beautiful garden in the northern India.[222]

Two gardens were laid down by Salabat Khan in the environs of the city of Ahmedabad. The gardens of *Farah Baksh* and *Bihisht bagh*, which presented a beautiful spectacle.

The garden of Farah Bakhsh is about 2000 zir'a in length and breadth alike, which made (the area) 278 bighas. In the middle of it is a reservoir 528 zir'a (square), which makes the area 19 bighas. An underground channel brings water to it from the foot of the hill. In the centre of the reservoir is a lofty and wonderful building in two stories, having 160 hujra (rooms) and a high cupola. Archer practice shooting at its summit. Bihist bagh is 312 zir'a in length and breadth alike which makes the area 100 bighas. In the middle is an octagonal reservoir, which is also fed with canal water. In the centre is a building now in ruins. On the bank of the reservoir are a charming building and a neat turkish bath, fit for the residence of elegant people.[223]

In the city of Farrukhabad[224] which was founded in 1723 by Mohammed Khan Bangash (1713-1793), after the name of reigning Mughal emperor Farrukhsiyar (1713-19), Mohammed Khan built a fort with three main gates all opening to East and within the fort, he laid the foundation of a mosque, a *Diwan-khana*, which was called *Bara mahal*, a palace and a garden. In the middle of the garden (later known as the garden of Begam Mukhtar Mahal) a *baradari* was situated.[225] There were several other gardens inside and outside the city of Farrukhabad. *Bahist bagh*, situated in the northwest corner and at a distance of about half a mile at western side at Neakpur *Khurd*, outside the walls, was the *Hayat bagh*. There were gardens like *Aish bagh*, *Pain bagh* and *Naulakha bagh* (a garden having reputedly 900,000 trees) all within the walls. Afterwards several gardens were laid out within the city at the instance of English East India Company servants'.[226]

There are numerous references to the gardens being laid down by the provincial rulers in the later period. With the invasion of Nadir Shah in 1739, and defeat of Mughals, the spell of Mughal power was broken and during the same year the Marathas invaded Delhi and led to the destruction of reputation of the Mughal army, and exhausted its accumulated treasure. With the result the history of gardens like paintings now passed on to provincial courts, the most prominent of which was the kingdom of Oudh, the eastern half of the present day Uttar Pradesh.[227] On the bank of the *Sarju*, about 80 miles from Lucknow is a little town engulfed among giant tamarind trees. It is Fyzabad which was once the capital of Oudh, which arose out of the shambles of the Mughal empire its founder was Sa'adat khan (1724 – 1739), the first *nawab wazir* of Oudh. He provided a new look to the town, a large number of gardens were laid out and shops sprang up there.[228] Later Nawab Shuja-ud-daula (1753-1775) laid down 5 famous gardens within the city. Those were *Anguri bagh*, *Motibagh*, *Lal bagh*, *Asaf bagh* and *Buland bagh*.[229] At Murshidabad also several gardens were laid down by the *nawabs* and their nobles and members of their families. The famous *Raushan bagh* was a garden tomb which contained the graves of Alivardi Khan and his family members. It was enclosed with a wall. *Farrah bagh* was originally a garden house of Nazir Ahmad, a district officer under Murshid Quli khan. It was confiscated by Shujauddin Khan wcho converted it into a pleasure garden. *Nishat bagh* was the residence of Muhammed Riza Khan the deputy diwan and deputy governor.[230]

In the 18th and 19th centuries, the European gardens especially the English gardens had a profound impact on the gardens in India. The British ruling class in India in the 18th century was largely derived from the landed aristocracy of Great Britain. Many of them had great interest in agriculture and gardening.[231] Warren Hasting, the British Governor General (1772–1782), was a scholar, a patron of learning and a keen gardener. Dennis Kincaid, describing his interest in horticulture, writes: *"He was almost happiest when pottering about a garden in the shabbiest of clothes."*[232] At Madras (1639 – 92) too, entertainments were often given in the gardens by the Britishers. In these gardens were grown all sorts of flowers, herbs for salads and fruit. In Pondicherry the French took initiative to grow parks having elegant geometrical designs, great square lawns with star shaped flower-beds and long well sweep walks protected from cows by creeper-hung pergolas. The English occasionally gave private parties in their gardens. In the shade of mango trees carpets were spread and cushions and mattresses, and mosquito nets were hung from branch to branch enclosing the party in an airy tent.[233]

The gardens created by the Mughals undoubtedly rank as one of the great landscape tradition of the world. Their characteristics are strong and their sense of design impeccable.[234] Mughal gardens in India can be divided into three categories according to their plans and surroundings. These are:

1) Mughal gardens in plains: for instance, *Char bagh* constructed by Babur in Agra, and the gardens at Shahjahanabad, Lahore etc.
2) Mughal gardens on hill slopes: A large number of gardens belonging to this category are situated in the province of Kashmir.
3) Then pattern of hill side gardens in the plains. Shahjahan adopted this method and the most fine example is *shalamar* garden Lahore.[235]

The garden tradition which Babur, the first great Mughal, was to carry to the hot dusty plains of northern India was purely Persian in concept, enclosed, private and quadripartite—the *Chahar or Char Bagh*. But over the next two centuries his descendants with their passion for garden building took this Persian form and transformed it into another which is unique in the garden tradition world wide.[236] The topography of northern India had to be considered as well, and particularly that of Kashmir. In the Perso-Arab tradition of building in a desert environment, every last drop of water was of physical as well as visual value, heightened by the privacy of the shaded gardens in which it was enjoyed. Away from the hot plains of India, in the cool valley of Kashmir water was not a premium, there were rushing mountain streams and lakes to delight the Mughal emperors. Water was abundant, for in addition to the two great lakes, there were several springs of great force and innumerable small streams. Unlike in Persia, the vegetation here was prolific and even at twelve thousand feet the land supported rhododendron and juniper. On the lower slopes are spruce, pines and cedars, while in the valley itself plane poplar and willow provided the principal shade trees, and the terrain had encouraged a deep and fertile soil, built up through centuries. Indeed, flat land suitable for garden building was scarce, as the mountain dropped rapidly into the lake, and ground levels and contours had to exploited by terracing. But water remained the unifying theme in both situations, its use in the narrow channels of the flat, quartered gardens of Persia was translated into ever increasing, rushing water displays of greater complexity in India, and it was this tradition which was later transformed back to water gardens first of the *Zands* and then of the *Qajars* in Persia.[237] Thus water was not only the basic form of design, but every detail of garden was related to the essential life stream of irrigation and to the delight which only water can give. In Kashmir spectacular natural water falls provided the prototypes for the Mughal cascades, while existing springs, such as those of Achhabal and Verinag already places of worship, became the focal points round which whole new gardens were developed. Mughals and Kashmiris shared a reverence for water, its quality, taste and coldness were keenly appreciated and discussed[238]

A more generous supply of water constantly inspired designers to new flights of imagination. The single jet from Persian tanks finally developed in to hundreds of fountains as in Shahjahan's *Shalamar* garden at Lahore, or the artificial rains of Udaipur.[239] The cascade crashes over the levels of Achhabal, mighty chute at Nishat bagh and in *Sahelion-ki-bari*, the use of falling water from the roof of a circular pavilion are the fine examples to show that water was used as decorative element in the gardens.[240] Also with abundance water in India the Persian pattern of narrow rills developed into wider canals and great tanks as the Mughals discovered the joys of cool air generated by the large sheets of water and the need for this in Indian heat. This development can be traced from the early Humayun's tomb at Delhi where the rills are still narrow on the Persian pattern, to the Taj where they have broadened out to 18 feet. In Kashmir the terrain, too offered new opportunities. The steep mountain sides scope for dramatic water landscapes. Powerful springs and streams, emerging from the hills fell with increased force over chutes and waterfalls or were disciplined into wide pools filled with fountains. Terrace succeeded terrace as for instance at *Nishat Bagh* on Dal lake, where there were no less than twelve terraces, the lowest one discharging its waterfall triumphantly into the lake.[241]

Order and symmetry were the keynote of the Mughal gardens. Babur brought to India, the love of order and symmetry that was equally a part of paradise garden tradition. Symmetry was the keynote of the Mughal art of gardening and guiding principle of Mughal gardens.[242] Almost all the Mughal gardens utilise the Persian principal of *char-bagh* i.e. a regular arrangement of squares, often subdivided into smaller squares to form the favourite figure of *char-bagh* or "four fold plot". Paved pathways and water channels follow the shape of these squares oblique and curved lines being rarely used.[243]

In its most basic form, the garden had four water channels, each crossing into the centre of the garden and creating four orderly quadrants. These channels were ordinarily placed over the level of the surrounding grounds so that their waters might fed the lines of trees planted along their banks as well as the fruits and flowers growing out within the quadrants and beyond and in the centre, at the meeting of four channels a *Baradari* (a canopied building with 12 open doors on sides. It is a typical Hindu structure) was often constructed to provide shade and rest amidst the cooling waters.[244] According to Mohibul Hasan:

Babur's gardens were in the form of terraces, each terrace being divided into four lesser squares according to the traditional plan of what is known as charbagh (four fold plot) with water channels for the distribution of water. Pavilions, supported by high columns occupied a central position. This was the pattern followed by the subsequent Mughal gardens.[245]

To ensure privacy it was the custom for the entire garden to be enclosed within a high wall which was a feature of the original paradise garden, and as indication of the considerable scale of some of these conceptions that at *Shalamar* near Lahore forms an oblong 1600 feet by 700 feet so that its longest measurement from end to end is over a third of a mile.[246] The Mughal gardens were mostly in the form of terraces. Not only in the hill gardens of Kashmir but this pattern was also followed in the flat plains. Sometimes there are eight terraces corresponding with the eight divisions of paradise, according to the Muslim faith, or seven to symbolise the seven planets. The *Nishat bagh* built by Asaf Khan, is in twelve terraces, one for each sign of the Zodiac.[247] At the central point in the garden scheme, masonry pavilions, loggias, kiosks and arbours were erected. Fro instance, there is a pillared pavilion of black marble in the middle of the *Shalamar bagh* in Kashmir, having no little architectural merit. To provide the water supply required to maintain such gardens in a state of uninterrupted efficiency it was often necessary to obtain this from a distant source by means of a canal, the construction of which was by no mean a work of engineering.[248] Chandra Bhan Brahaman writes:

... beautiful buildings, flowing water, and a number of tanks, canals with greenery and flowers were the important elements of the Mughal gardens in Kashmir[249]

Faced with the paradise of Kashmir the remarkable achievement of the Mughals was that, without having the strength of their traditional design, they responded to the new conditions. The strength of the mountains as a defence was accepted, the garden walls modified from the complete barrier of the Persian originals to allow landscape and gardens to drift into each other[250] The Mughals understood the very roots of garden design. They combined strength of composition with a feeling for the site, and created gardens for the pleasure of living, in whatever place and climate they found themselves. They knew how to make the best use of restricted space when necessary. Their choice of material and workmanship was superb. They understood and used their native flora, creating gardens which were a reflection of both their way of life and the conditions of their country.[251]

Mughals were the great lovers of beauty and in fact they regarded gardens as important part of their social life, which served them as pleasure resorts as long as they lived, they also used them as their burial ground when they died. They usually constructed their tombs in the middle of their gardens with great "magnificence", during their life time.[252] There are numerous instances which show that Mughals used to celebrate the important occasions of their lives as well as their festivals in the gardens. Gulbadan *Begum* writes that the royal ladies celebrated the Babur's victory at Panipat in the *Bagh-i-diwan khana*.

Tents with the screen were set up in the garden of the Audience hall for each begum. Three days they remained together in the Audience hall garden. They were uplifted by pride, and recited the fatiha for the benediction and prosperity of His Majesty (Babur) and joyfully made the prostration of thanks.[253]

In the year 1534, the marriage of Hindal Mirza, brother of Humayun was solemnised and the feats and celebrations of his marriage were organised in the royal garden. Khwanda Mir writes:

The marriage of...Nasr Muhammed Hindal Mirza was also to be solemnised, merriment and pleasure were greater than on any occasion previously...The nobles and the ministers built chahar Taqs[254] in the royal garden, and adorned them with fine clothes and other magnificent articles.

Verses:- On all sides in the paradise like garden,
were erected more chaharTaqs
They were adorned with Turkish and European clothes
and with embroided and seven coloured clothes.[255]

Gulbadan *Begum* describes the celebration of Akbar's circumcision in *bagh-i-Diwankhana* in the year 1545.

A few days later he (Humayun) sent persons to bring Hamida Banu Begum from Qandhar. When she arrived they celebrated the feast of the circumcision of the emperor Jalaluddin Muhammad Akbar. The emperor Muhammad Akbar was five years old when they made circumcision feast in Kabul. They gave it in that same large Audience hall garden (bagh-i-diwankhana). All the sultans and amirs brought gifts to the Audience hall garden. There were many elegant festivities and grand entertainments, and costly Khi'lats and head to foot dresses were bestowed.[256]

The gardens were also the excursion grounds for the Mughal emperors. Gulbadan *begum* writes about Babur's excursion to the Gold scattering garden (*Bagh-i-zafran*) at Agra.

A few days later he (Babur) made an excursion to the Gold scattering garden (bagh-i-zarafshan) (Agra). There was a place in it for ablution before prayers. When he saw it, he said: 'My heart is bowed down by ruling and reigning; I will retire to this garden. As for attendance, Tahir the ewer-bearer will amply suffice. I will make over the kingdom to Humayun.[257]

Similarly *Begum ka bagh* at Delhi was frequently visited by the ladies of the harem and wives and daughters of the nobles frequently. They spent hours of peaceful enjoyment there, punctuated by merriment and gaiety on the many swings provided. The garden was also a venue for many festivals, not able among which was the *Pankhon ka mela* This *mela* was essentially for ladies. It was celebrated for seven days inside the red fort and on the eighth day, royalty and nobility came in procession to a *dargah* near *Begum ka bagh*.[258] During the reign of Shahjahan the festival of *Nauroz* was celebrated in the gardens. Inayat Khan writes:

The nauroz festival of this happy omen year fell on the 21st of Ramazan 1043 (21 March 1634), and was celebrated in the Bagh-i-Hafiz Rakna at Sirhind.[259]

A traditional secular fair called *Phool walon ki sair* (festival of flower sellers) was celebrated at gardens in Shahjahanabad.[260] Similarly *Id-i-Milad* or the feast of Prophets was celebrated with great solemnity and on that day palaces, gardens and reservoirs were all illuminated.[261] There are references to the celebration of the festival of *Basant* in gardens.

of all such festivals, Basant was the most popular, with both Hindus and Muslims participating. It was celebrated continuously for seven days and nights.[262]

Mughals had so deep love for nature that every place they visited whether for war purpose or for any other they preferred to stay in the gardens and there are numerous references to prove it. When Babur came to India, he made his halt at the garden of Mirza Kamran at Lahore and Mirza Kamran gave a magnificent banquet there which lasted for 3 days.[263] In *Baburnama* there are numerous references which show that both Babur and Humayun preferred to stay at gardens during their expeditions. It has been narrated in *Baburnama* that Humayun after completing his expedition in Eastern India in 1527 went to Agra and there he waited for Babur in the Garden of Eight Paradises on 6th January 1527.²⁶⁴ There is also the reference to Humayun's stay at Gold Scattering Garden about 6 miles from Agra on 20th Jan.1527.²⁶⁵ *Baburnmma* gives numerous references to Babur's halt at various gardens during his expeditions in India. For instance, he encamped at Lotus garden at Dholpur on 9th Jan.1527.²⁶⁶ Like his father Humayun also preferred to stay in gardens during his expeditions. Abul Fazl has mentioned in *Akbarnama* that in 1540 when Humayun was defeated by Sher Shah, he marched towards Lahore, there he stayed in the garden of Khwaja Dost Munshi, while Mirza Hindal took up his quarters in the garden of Khwaja Ghaji, who was then Mirza Kamran's Diwan.²⁶⁷ Similarly on his way to Kashmir Shahjahan in 1645 encamped in the pleasant and agreeable gardens of Faiz Baksh and Farah Baksh at Lahore.²⁶⁸ Aurangzeb also preferred to encamp at gardens during his expeditions. The coronation of Aurangzeb also took place in the garden of *Shalamar* at Delhi in 1658.²⁶⁹

Mughals were so much fond of gardens that they used to make their tombs in the gardens. Babur's body was laid first in the Ram bagh (Garden of rest), Agra on the opposite side of the river from the present Taj-Mahal. Later it was taken to Kabul and buried at *bagh-i-wafa*. Humayun's tomb was made in 1573 at Delhi.²⁷⁰ After her death Mumtaz Mahal was buried in the garden of Zainabad at Burhanpur on the other side of river Tapti. Later on her body was shifted to Agra and buried in the most magnificent garden tomb.²⁷¹ Empress Nurjahan was also buried in a garden close to the tombs of her husband and brother.²⁷² Princes and nobles built their own tombs during their lifetime only, the surrounding gardens were used for the purpose. These tomb gardens embodied the Muslim ideal of paradise to be enjoyed by both the living and the dead. In Quran it has been quoted:

> *With O' erbranching trees in each:*
> *In each two fountains flowing:*
> *In each two kinds of every fruit:*
> *On couches with linings of brocade shall they recline,*
> *And the fruit of the two gardens shall be within easy reach:*

> *There in shall be the damsels with retiring glanus,*
> *Whom nor man nor djinn hath touched before them:*
> *Like Jacinths and pearls:*
> *Shall the reward of good be ought but good?*
> *And besides these shall be two other gardens:*
> *Of a dark green:*
> *with gushing fountains in each:*
> *In each fruits and the palm and the pomegranate:*
> *In each the fair, the beauteous ones:*
> *with large dark eye balls, kept close in their pavilions:*
> *Whom man has never touched, nor any djinn.*
> *Their spouses on soft green cushions and on beautiful carpets shall recline:*
> *Blessed be the name of thy Lord, full of majesty and glory.*

The builders of the tombs would spend whole days with their families and court in the surrounding gardens. this passion for living in the open influenced the design of their gardens.[273] Mughals established a regular department for the supervision of the gardens and orchards.[274] A large amount of money was spent in the maintenance of these gardens. Shahjahan spent 25 millions of rupees on the erection of gardens and other buildings at Agra, Delhi, Lahore, etc.[275] Jahangir was so devoted to the gardens that to encourage the plantation of gardens he exempted land revenue of those lands which were converted into orchards.[276]

Gardens constituted a very special place in the life of Mughals. The innumerable poems about gardens, and description of gardens in the Mughal poetry, all testify to the extent to which the Mughal rulers and their subjects loved gardens. Above all others, it is the roses which feature in the verses by the Mughal poets, for their beauty in gardens symbolises fleeting happiness:

> *The same in shape and form are*
> *Joy and pain–the rose*
> *Call it an open heart…*
> *Call it a broken heart.*

This was written by Mir Dard. His contemporary Azad Bilgrami also viewed gardens as symbols of the impermanence of life.

Only with age do we appreciate
Scent and colour
The young bud:ignorant,
And still blind…[277]

REFERENCES AND NOTES

1. Randhawa. *Gardens through the Ages*, P-9.
2. Boorks, John. *Gardens of Paradise* (The History and Design of Great Islamic Gardens). London: George Weidenfeld and Nicolson Ltd., 1987.p-17.
3. William Bridgwater and Seymour (ed.) *The Columbia Encyclopedia*, 3rd ed. p-794.
4. Shadoof is similar to the dheengli of India, a counter poised beam used for raising water by leverage. Randhawa. *Gardens through the Ages*, p-40.
5. Coomarswamy, Ananda K. & sister Nivedita. *Myths of Hindus and Buddhists*. Mumbai: Jaico publishing House, 1999. p-19. These forests were called hermitages or *ashramas* where pupil attained knowledge.
6. Basham, A.L. *The Wonder that was India*. Calcutta: Rupa&co.,1996. pp-202, 203.
7. Randhawa. *Gardens through the Ages*. P-17.
8. Naqvi. *Urbanisation and Urban Centres*, p-1.
9. Ansari. *Geographical Glimpses*, Vol.2, Intro. XV.
10. Afif. *Tarik-i-Firozshahi* in *Geographical Glimpses* Vol.2, p-8, 10.
11. Chittor was called Mewar. In 1595 conforming to the limits of that principality, most of its paraganas were under district control and assigned as *jagir* to Imperial Nobles. Habib, Irfan. *An Atlas of Mughal Empire*, p-57, sheet 14A.
12. Afif. *Tarikh-i-Firozshahi* in *Geographical Glimpses*, Vol.2, p-10. Also see R.C.Jauhari. *Firozshah Tughlaq*, pp-107, 108.
13. Chandra, Satish. *Medieval India from Sultanate to Mughals*. Part-I, Delhi: 1999.p-234.
14. Ashraf, K.M. *Life and Conditions of the People of Hindustan*. New Delhi: Munshiram Manohar Lal,1988. pp-119, 120.
15. Ahmad, Khwaja Nizamuddin. *TheTabqat-i-Akbari* vol.I Trans. by Brajendranath. Jammu: Jay Kay Book House,1994 p-276.
16. Bendrey, V.S. *A Study of Muslim Inscriptions*. Delhi: Anmol Publications, 1985. p-112.
17. Brown, Percy. *Indian Architecture* (Islamic Period). Bombay: D.B.Taraporevala sons & co.,1995. p-109.
18. Ansari, *Social Life of the Mughal Emperors*, p-56.
19. Chaudari, Tapan Ray and Irfan Habib (ed.) *The Cambridge Economic History of India*. Vol.I. C.1200-1750. New Delhi: Oriental Logman in association with Cambridge University Press.P-
20. Khan, Zain. *Tabqat-i-Baburi*. Trans. and introduction by Sayed Hasan Askari. Delhi:Idarah-i-Adabiyat-i Delli,1982. p- intro XVII.
21. i.e. Shahruki, Timur's grandson; *Baburnama*, p-77.

22. *Baburnama*, pp-77, 78.
23. *Ibid*, p-80.
23a. Schimmel, Annemarie. The Empire of the Great Mughals. reprint London: Reaktion Books Ltd.,2010 p-295
24. Gulbadan Begam. *Humayun-nama*, p-22.
24a. The Empire of the Great Mughals.Op.cit. p-295
25. *Baburnama*, p-208.
26. *Ibid*, pp-208, 209.
27. Iram – built by Shahdad, the name of a impious king, the founder of the garden of Iram. C.S.Steingass, Persian-eng.dictionary, f.t. 12a in Zainkhan's *Tabqat-i-Baburi*, p-183.
28. *Baburnama*, appendices pp-Ixxix-Ixxx.
29. Sylvia Crowe, p-62.
30. *Baburnama*, p-487.
31. *Ibid*, p-519.
32. Hasan, Mohibul. *Babur.* New Delhi: Manohar Publishers, 1985 p-137.
33. *Baburnama*, p-531.
34. *Ain* Vol.1, p-933.
35. *Baburnama*, pp-531, 532.
36. Zainkhan.*Tabqat-i-Baburi*, p-159.
37. Sylvia Crowe, P-63.
38. *Tuzuk*, Vol.1, pp-4, 5.
39. Mohibul Hasan, *Babur* p-197.
40. *Baburnama*, p-640.
41. Gulbadan Begam. *Humayunama*, pp-103-108.
42. Biana or Bayana was the Pargana situated in the Sarkar Agra of Suba Agra. Irfan Habib. *An Atlas of Mughal Empire*, p-19, sheet-6A.
43. *Baburnama*, p-581.
44. *Chaukandi*, an open four side pavilion, either a free standing structure or a pavilion mounted on an elephant. *Turkhana* perhaps a space enclosed by a low railing. Possibly and suitably a mosquito room. Gulbadan Begum's *Humayunnama*, p-103
45. *Ibid*, p-581.
46. *Ibid*, p-644.
47. Sylvia Crowe, p-60.
48. *Baburnama*, p-710.
49. Allami, Abul Fazl. *The Akbarnama*, vols.I, II &III. Trans. H.Beveridge. reprint Delhi: Low Price Publications, 2017. vol.I, p-362.
50. Gulbadan Begam, *Humayunama*, p-148.
51. *Tabqat-i-Akbari* of Nizamuddin in Elliot & Dowson, Vol.5, p-218.
52. *The Akbarnama* Vol.I, pp-355, 356.

53. D. Pant, *The Commercial Policy of Mughals*, P-32.
54. Khwandamir. *Qanun-i-Humayuni* or *Humayun-nama*. Trans. by Baini Prashad. reprint Calcutta: The Asiatic Society, 1996 pp-44,45
54a. Ibid, p-45
55. *Humayunnama* by Khondamir in Elliot & Dowson, Vol.5, p-124.
56. Father Monserrate, p-97.
57. Sylvia Crowe, pp-71-73, also see John Brooks, pp-131, 132.
58. Father Moneserrate, p-95.
59. Elliot and Dowson, Vol.5, p-487.
60. Sylvia Crowe, p-74.
61. Qandhari, Muhammed Arif. *Tarikh-i-Akbari*. Trans. by Tasneem Ahmad, Delhi: Pragati Publications, 1993. p-85.
62. *The Jahangirnama*, pp-5,6.
63. Sylvia Crowe, p-76.
64. Manucci, Vol.I, pp-137, 138.
65. Rajauri or Rajaur was located on the routes to Kashmir via Punchas, it was the state on route to Kashmir therefore it must have also belonged to Kashmir. Irfan Habib, an Atlas of Mughal empire, p-6, sheet 3-A.
66. Sylvia Crowe, p-80.
67. Stuart, C.M.Villiars, *Gardens of the Great Mughals*. London: A.and C. Black, 1913. p-155.
68. Sylvia Crowe, p-83.
69. *Tuzuk*, Vol.2, pp.150-151.
70. Villiars Stuart, *op.cit.*, p-158.
71. Fujiwara, Shinya. *The beautiful world, Kashmir* Vol.60. Delhi: Allied Publishers, 1982 p-44.
72. Sylvia Crowe, p-84.
73. Tirmizi, S.A.I. *Mughal documents (1526-1627)*. New Delhi: Manohar Publishers, 1989. p-56.
74. Sir Wolseley Haig, *The Cambridge History*, Vol.4, pp-178, 179.
75. John Brooks, pp-137, 138.
76. *Tuzuk*, Vol.I, p-106. Also See *The Jahangirnama*, p-76.
77. *The Jahangirnama*, p-76.
78. *Aurangzeb in Kashmir*; p-2.
79. *Tuzuk*, Vol.2, pp-143, 144.
80. John Brooks, pp-141, 143.
81. Bernier, p-399.
82. Sylvia Crowe, p-98.
83. *The Jahangirnama*, p-335.
84. Sylvia Crowe, pp-96, 102.
85. John Brooks, p-143.

86. *Ibid;* p-146.
87. *The Jahangirnama,* p-345.
88. Randhawa, *Garden Through Ages,* p-122.
89. Bernier, p-413.
90. *Ibid,* p-413, f.n.2
91. *Tuzuk,* Vol.I, p-92.
92. *The Jahangirnama,* p-346.
93. Bernier, pp-413, 414.
94. Sylvia Crowe, p-112.
95. Matto, *Kashmir Under Mughals,* p-201.
96. Findly, Ellison Banks. *Nurjahan, Empress of Mughal India.* New Delhi: Oxford University Press, 1993 p-255.
97. *Ibid,* p-255. Also see Chandra Pant, p-117.
98. Matto. *Kashmir Under Mughals,* p-197.
99. *The Jahangirnama,* p-345.
100. Pelsaert, p-50.
101. *Tuzuk* Vol.2, pp-75, 76.
102. Ellison. *Nurjahan,* P-129.
103. *The Jahangirnama,* p-355.
104. Ellison. *Nurjahan,* p-250.
105. Sylvia Crowe, p-123.
106. *Ibid,* p-131.
107. Rawalpindi occupies the northwest corner of the Indus plain area. The principal hills are the Salt Range. Imperial Gazetteer of India, the Indian Empire Vol.I, p-179.
108. Randhawa. *Gardens Through Ages,* p-121.
109. Sylvia Crowe, p-128.
110. *Tuzuk,* Vol.I, p-152.
111. *Ibid,* p-136.
112. Pelsaert, p-30.
113. *Ibid,* p-5.
114. William Foster. *Early Travels,* p-158.
115. *Tuzuk* Vol2, p-64.
116. John Brooks, p-151.
117. Richards, John F. *The New Cambridge History of India,* The Mughal Empire. Reprint Delhi: Cambridge University Press, 2011 p-139.
118. *The Shahjahan-nama,* pp-8, 9.
119. *Ibid,* p-126.
120. *Ibid,* pp-137, 138.
121. Shinya Fujiwara. *Kashmir* Vol.60, p-44.
122. Sylvia Crowe, p-138.

123. Bamzai, P.N.K. *A History of Kashmir: Political, social, Cultural*, from the earliest times to present day. Delhi: Metropolitan Book Co., 1962 p-369.
124. Randhawa, *Gardens Through Ages*, pp-124, 125.
125. Sylvia Crowe, p-140.
126. John Brooks, p-157.
127. Sylvia Crowe, p-136.
128. Percy Brown, p-83.
129. Sylvia Crowe, p-136.
130. Koul, Pandit Anand. *Archaeological Remains in Kashmir.* Lahore: The Mercantile Press, 1935. p-93.
131. Villiers Stuart, p-181.
132. Sylvia Crowe, pp-135, 136.
132a. *The Empire of the Great Mughals*, p-297
133. Shahjahanabad – In 1648 Shajahan built Red Fort at Delhi, he named his new city as Shahjahanabad, henceforth the Suba of Delhi too began to be called Shajahanabad.
134. John Brooks, p-155.
135. Silvia Crowe, p-157.
136. *Khulasat* in *Geographical Glimpses*, Vol.3, p-66.
137. Sylvia Crowe, p-160.
138. Maheshwar Dayal, *Rediscovering Delhi*, p-5.
139. Sylvia Crowe, p-159.
140. Manucci, Vol.I, p-178.
141. Maheshwar Dayal. *op.cit.*, pp-29 to 32.
142. *Ibid*, pp-32, 33.
143. *Ibid*, p-71.
144. *Ibid*, p-60.
145. *Ibid*, p-69.
146. Sylvia Crowe, p-146.
147. *The Shahjahan-nama*, p-451.
148. Maheshwar Dayal. op.cit., p-70.
149. Sylvia Crowe, p-147.
150. Maheshwar Dayal. Rediscovering Delhi, p-70.
151. *Chahar Chaman* by Chandra Bhan Brahman in Geographical Glimpses, Vol.3, p-5.
152. *Ibid*, p-6.
153. Percy Brown; p-110.
154. *The Shahjahan-nama*, p-277.
155. Sylvia Crowe, p-148.
156. *The Shahjahan-nama*, p-298.
157. Percy Brown, p-110.

158. Sylvia Crowe, p-150.
159. *Khulasat*, p-82.
160. Sylvia Crowe, p-155.
161. *Ibid*, pp-162-164.
162. Manucci, Vol.2, p-43.
163. *The Shahjahan-nama*, p-364.
164. *Ibid*, p-394.
165. Percy Brown, pp-107, 108.
166. Randhawa, *Gardens through Ages*, P-125.
167. Sylvia Crowe, p-170.
168. Percy Brown, p-108.
169. Bernier, p-295.
170. *The Shahjahan-nama*, pp-299, 300.
171. Sirhind, the reference is probably the garden of Hafiz Rakhnah. He was a grandee of Humayun's court (Ain II, 281). Both the mosque and park are mentioned by Monserrate (pp-101-102) and Manrique. De Laet, p-49, f.n.65.
172. *Ibid*, pp-49, 50.
173. Nagar, Ishwar Das. *Futuhat-i-Alamgiri*. Trans. by Tasneem Ahmad. Delhi: Idarah-i Adabiyat-i Delli, 1978 p-29.
174. Sylvia Crowe, p-176.
175. Khan, Saqi Must'ad. *Maasir-i-Alamgiri*. A History of Emperor Aurangzeb Alamgir. Trans. by Jadunath Sarkar. 2nd edition. New Delhi: Munshiram Manohar Lal Publishers, 1986 p-4.
176. Sylvia Crowe, p-176.
177. John Brooks, p-160.
178. Randhawa. *Gardens through Ages*, pp-128, 129.
179. Sylvia Crowe, p-183.
180. Tavernier, Vol.I, p-94.
181. Manucci, Vol.2, p-174.
182. Sylvia Crowe, p-184.
183. Maheshwar Dayal, p-231.
184. *Ibid*, pp-200-201.
185. Irvine, William. *Later Mughals* 1719-1739 Vol.2.Calcutta: M.C.Sarkar & sons, 1922
186. Randhawa. *Gardens through Ages*, p-139.
187. Verma. *News Letters of Mughal Court*, pp-1, 3.
188. *Baburnama*; p-532.
189. *The Tarikh-i-Rashidi*, p-424.
190. Sir Wolseley Haig. *The Cambridge History* Vol.4, p-89.
191. Nizamuddin Ahmad. *Tabqat-i-Akbari* in Elliot & Dowson, vol.5, pp- 406, 407.

192. William Finch. *Early Travels*, pp-138, 139.
193. *Ibid*, p-207.
194. *Ibid*, p-207, f.n. 1
195. *Tuzuk* Vol.I, p-429.
196. Aurangabad and Ahmadnagar – The Suba of Aurangabad broadly represented the former kingdom of Nizamshahs. The capital of the Nizamshahi Kingdom had been Ahmadnagar, the Ain, I, 386 refers to Ahmadnagar among the 3 newly acquired subas, the other 2 being Berar and Khandesh in Deccan. Although the annexation of the kingdom of Ahmadnagar had been decreed by Akbar in the 40th regnal year (1595-96) (Ain III, 698) but the real annexation of the Kingdom took place in 1633 with the seizure of Daultabad (Lahori, Vol.I, 530-31). The Suba previously designated Deccan (Dakin) i.e. Ahmadnagar. But the records of Aurangzeb's reign onwards style it as the suba of Aurangabad, after the name of his new capital. Irfan Habinb, An Atlas of Mughal Empire, p-54, sheet-14A.
197. Noorudin's *Latifa-i-Faiyazi* in Ansari's Geographical Glimpses, Vol.2, p-24.
198. Ami Ahmad Razi's *Haft Iqlim* in Ansari's Geographical Glimpses, Vol.2, p-21.
199. Father Monserrate, p-219.
200. Ibid, p-102.
201. Mohammad Amin Qazvini's *Badshah-nama*, in Ansari's Geographical Glimpses, Vol.3, p-1.
202. Nizamuddin, *Tabqat-i-Akbari*, Vol.1, p- Preface XV.
203. Noorudin's *Latifa-i-Faiyazi* in Ansari's Geographical glimpses. Vol.2, p-24.
204. Gulberga was situated in the Deccan West. It was called Ahsanabad, once the seat of Bahmani Kings. Irfan Habib, An Atlas of Mughal Empire, p-57, sheet 14-A.
205. William Foster, *Early Travels*, p-1.
206. Bendrey, *Muslim Inscriptions*, p-133.
207. *Tuzuk*, vol.2, p-112.
208 Chandra, Pant. *Nurjahan and Her family*, p-16
209. William Foster, *Early Travels*, p-165.
210. John Brooks, p-143
211. Chandra Pant, *Nurjahan*, p-147
212. Randhawa; *Gardens through Ages*; p-122
213. Chandra Pant, *Nurjahan*, p-148.
214. *The Shahjahan-nama*, p-126.
215. *Ibid*, p-127.
216. Sodhra is about 4 miles north-east of Wazirabad in Pakistan, *Khulasat*, p-98, f.n.1.

217. Ali Mardan Khan's Canal "Brought the waters of the Tavi to supply the Imperial gardens at Shahdara", on the Ravi, opposite Lahore city. *Khulasat*, p-98, f.n.2.
218. *Khulasat*, p-98.
219 *Journal of Pakistan Historical Society* part-I, Vol.XXXVIII, p-348.
220. *Gardens through Ages*, p-129.
221. *Khulasat in Geographical Glimpses*, Vol.3, p-68.
222. *Gardens through Ages*, p-129.
223. *Maasir-i-Alamgiri*, p-157.
224. Farrukhabad town, U.P. State, N.India, a joint Municipality with Fategarh. See *The Columbia Encyclopedia*, 3rd Ed.
225. Muhammed Umar. *Urban Culture in Northern India during the 18th C.* Aligarh: 2001, p-23.
226. *Ibid*, p-24, 25.
227. Randhawa. *Gardens through Ages*, pp14-, 141.
228. *Ibid*, p-142.
229. Muhammed Umar. *Urban Culture*, p-4.
230 Muhammed Umar. *Urban Culture*, p18.
231. Randhawa. *Gardens Through Ages*, p-261, 262.
232. Dennis Kincaid. *British social Life in India*, pp-110, 118.
233. *Ibid*, pp-62 to 64.
234. Sylvia Crowe, p-44.
235. Ansari. *Social Life of the Mughal Emperors*, pp-56 to 60.
236. John Brooks, p-116.
237. *Ibid*, p-119, 122.
238. Sylvia Crowe, p-44.
239. *Ibid*, p-45.
240. John Brooks, pp-194, 195.
241. Sylvia Crowe, pp-44, 45.
242. Jaffar, S.M. *Some Cultural Aspects of Muslim Rule in India.* Delhi: Idarah-i Adabiyat-i Delli, 1972 p-122.
243. Percy Brown, p-110.
244. Ellison. *Nurjahan*, p-246.
245. Mohibul Hasan. *Babar*, p-196.
246. Percy Brown, p-110.
247. Randhawa. *Gardens Through Ages*, p-134.
248. Percy Brown, p-110.
249. *Charhar Chaman* of Chander Bhan Brahaman, in Geographical Glimpses, Vol.3, p-6.
250. Sylvia Crowe, p-47.
251 *Ibid*, p-53.

252. Ansari. *European Travelers*, p-126.
253. Gulbadan Begum. *Humayunnama*, pp-95,96
254. *ChaharTaqs* Mrs. Beveridge (memoirs of Babur, p-264, note1,1914) explains *chaharTaqs* as a large tent rising into four domes or have four porches. It was some kind of a tent used during Humayun's reign.
255. Khwandamir. *Qanun-i-Humayuni*. p-64
256. Gulbadan Begum. *Humayunnama*, pp179,80
257. Ibid, p-103
258. Maeshwar Dayal. *Rediscovering Delhi, p-31*
259. *The Shahjahan-nama*, p-122.
260. Maheshwar Dayal, p-200.
261. Prannath Chopra, *Some aspects of Society and Culture during Mughal Age (1526-1707)*. Agra: Educational Publishers,1955 pp-101, 102.
262 Saiyid Athar Abbas Rizvi, *Shah Wali Allah and His Times*, p-176.
263. Ahmad Yadgar's *Tarikh-i-Afghana* in *The History of India as told by its own Historians*, Elliot & Dowson, Vol.5, p-40.
264. This is the Charbagh, known later as the Ram [Aram] Bagh (Garden of Rest). *Babur-nama*, p-544.
265. *Ibid*, p-640.
266. *Ibid*, p-639.
267 *The Akbarnama*, Vol.1, p-355.
268 *The Shahjahan-nama*, p-323.
269. *Maasir-i-Alamgiri*, p-4.
270 Sylvia Crowe, p-71.
271 *The Shahjahan-nama*, p-70.
272. Sir Wolsely Haig. *The Cambridge History of India*, Vol.4, p-202.
273. Sylvia Crowe, pp-42,44
274. Naqvi. *Urbanisation and Urban Centres*, p-104.
275 John F.Richards. *The New Cambridge History of India*, p-139.
276. Compare, Irfan Habib, *The Agrarian system of Mughal India*, p-49.
277. *The Empire of the Great Mughals*. p-297

SOCIETY AND PLANTATION OF TREES

Human beings' dependence on the trees began during the palaeolithic period. Even before man developed the agriculture, he was dependent on trees for fulfilling his needs, as he lived mostly on fruits and flesh. Trees provided him shelter against inclement weather, fruits and nuts to eat, wood for implements. It was from wood that he obtained fire which enabled him to cook food and to warm his dwellings. As Fergusson observes:

With all their poetry, and all their usefulness, we can hardly feel astonished that the primitive races of mankind should have considered trees the choicest gifts of the Gods to men, and should have believed that their spirits still dwelt among their branches, or spoke oracles through the rustling of their leaves.[1]

And with the development of civilization the need for the tree products increased. Aryans had great love for trees and flowers. They have praised the forests filled with flowering trees. From 800–600 BC books on forests called *Arayankas* i.e. "forests books" were written. In these they have revealed the luxuriance of Indian trees and flowers.[1a] In *Mahabharata* also there are numerous references to several trees such as mango, *asoka*, *champaka*, *nag-champa*, *sal*, palmyra, skrew pine bignonia, coral tree and oleander. In Mahabharata there is also find reference to *Kadamba* trees in Kamyak forest which existed south West of Delhi and it is likely that the existing *Kadamba* forests in Mathura and Bharatpore are remnants of this ancient forests.[2] In ancient India, in every village, the planting of banyan and *pipal* tree was enjoined. Apart from shade, it was also a measure for saving crops and fruits from destructive birds. Banyan and *pipal* trees, when covered with figs, provided food to thousands of birds. Thus indirectly they saved fruit trees from the damage by birds which were kept busy eating their figs for weeks.[3] During Mughal period more and more emphasis was laid on the plantation of more and more trees as trees were associated with all the socio-economic activities of the people and furnished them with all their needs.

Tree plantation was not only encouraged by the ruling class but also by the people in general. All section of society took great effort in tree plantation. Smaller plots of land available within or on the outskirts of the towns which were not conveniently situated for regular cultivation were used for the plantation of the flower and fruit trees. Many large orchards with numerous fruit trees were planted between towns and villages by well off urban citizens and the rich peasants who could with little difficulty manage to plant orchards side by side and gain additional income.[4] It was general practice among the Muslims to build the graves among the groves of the fruit trees. Trees bearing the better class fruits such as quality mangoes were usually planted in the groves, in carefully measured rows.[5] The assignment of *madad-i-maash* lands to the needy and pious and to intellectual persons was a religious requirement of the state under Islamic law. Under Mughals this practice received greater attention. The *Madad-i-Maash* holders planted orchards on the large scale. They were given revenue free lands specifically for this purpose. In Awadh the grant holders planted and supervised the growth of mango, guava, and other fruits. These groves helped to strengthen the rural economy, and also provided some relief to the people during summer season. It may be mentioned that the fruits from the orchards were not only consumed by the owners, but these were also sold for the profit. Since orchards were a source of income, it seems that the grantees might have been encouraged to grow trees in large numbers. Thus with the plantation of the orchards, *madad-i-maash* holders improved their economic condition and also provided fruit to the locality.[6] In Awadh gentry and land holders also encouraged the cultivation of the fruit trees. It was the Muslim service gentry of the small towns and large villages of Awadh who planted and supervised the growth of guava, mango and other fruit trees planted around their home market towns. Specialist market gardeners and fruiterer castes *(malis, kacchis, and koeris)* were given leases of groves, but the gentry retained an interest in improvement and marketing.[7] Sufi saints also played a very important role in the plantation of trees in many regions. Jammu which is situated between Punjab and Kashmir attracted the attention of sufis from both the directions. Pir Roshan Wali Shah is believed to be the earliest sufi who came to Jammu from Mecca in the first half of the thirteenth century. But the hagiographical traditions inform us that the majority of sufis arrived in Jammu from the fifteenth to nineteenth century. Pir Lakhdatta, Pir Buddhan Ali Shah, Hazrat Zainuddin Rishi, Baba Latifuddin Rishi, Pir Mitha belonged to the fifteenth century. But a large number of sufis came from Punjab during the eighteenth and nineteenth centuries. Mention may be made of Baba Jiwan Shah, Mustafa or Nau-Gaza Baba. These sufi saints apart from maintaining and promoting social harmony, the sufis of Jammu hills made an earnest effort to maintain ecological balance. For instance, they participated in the plantation of trees and founded water resources particularly the *baolis* (wells with steps). it was believed that leaves and fruits from the trees planted by them and the water of these resources was helpful in curing certain diseases. Shah Gulam Badshah of Rajouri is said to have planted a tree which became everlasting one known as *Sadabahar*. It still survives and yields fruit through the year, but the plucking of fruit is forbidden. Only that fruit is used which falls on the ground. It is believed that whosoever gets the fruit and eats it, his or her prayers are answered. The uniqueness of the fruit of this tree is that it never gets perished. It is also believed that whosoever wishes to be blessed with a child must eat the leaves of this tree. The concern of the Jammu sufis with ecology led to the evolution of the culture of preserving trees and clean water. A large number of rites in this region are associated with water resources and trees.[7a]

In Kashmir too the culture of tree plantation by the sufi saints prevailed. Abul Fazl in his *Ain-i-Akbari* writes:
The most respectable class in this country (Kashmir) is that of the rishis, who not withstanding their need of freedom from the bonds of tradition

and custom, are true worshippers of God. They do not loosen the tongue of calumny against those not in their faith, nor beg nor importune. They employ themselves in planting fruit trees and are generally source of benefit to the people. They abstain from flesh and do not marry. There are about 2000 of this class.[7b]

In his *Tuzuk* Jahangir too makes a reference to the plantation of fruit trees by the *rishis* of Kashmir for the benefit of the common people.

There is also a body of Faqirs whom they call rishis. Though they have not religious knowledge or learning of any sort, yet they posses simplicity, and are without pretence. They abuse no one, they restrain the tongue of desire, and the foot of seeking; they eat no flesh, they have no wives and always plant fruit-bearing trees in the fields, so that men may benefit from them, themselves driving no advantage. there are about 2000 of them.[7c]

Common people also encouraged the tree plantation in their houses. The Bengal houses were conspicuous for the construction of a tank, an orchard in which numerous trees were grown and a bamboo grove. The houses in Orissa also consisted of the fruit trees. Houses had orchards of fruit trees in Cambay which was 'a most excellent city'. People of Khambayat had 'many vegetable and fruit gardens and orchards which they used for pleasures.' Marwari merchants of those days constructed many water tanks in their houses in addition to the usual orchards and gardens.[8] Apart from Muslims, Hindus also encouraged tree plantation. There were the groves of palm and mango trees planted all over the Empire – the former in the vicinity of the coats, the latter in the north-west provinces and Bihar. A strong religious feeling influenced the Hindus in these plantations. They believed that their soul in the next world would be benefited by the blessings and grateful feelings of those of their fellow creatures who, unmolested, eat the fruit and enjoy the shade of the trees, they had planted during their sojourn in this world.[9] Tree plantation was encouraged in all parts of the Empire for instance Ralph Fitch has described that the city of Goa was full of orchards and there were numerous Palm trees.[10] Pelsaert has also mentioned that the trees were plentiful round the cities.[11] Bernier writes about the plantation of large number of trees in Kashmir. He writes:

The lake is full of islands, which are so many pleasure grounds. They look beautiful and green in the midst of water, being covered with fruit trees, and laid out with regular trellised walks. In general they are as high as the mast of a ship, and have only a tuff of branches at the top, like the palm trees.[12]

People also encouraged the plantation of trees along the roadsides. A beautiful description of the tree plantation the roadsides of India has been given by Sujan Rai:

The roads are two ways with trees lining all along. Trees are fruit bearing... Travellers travel in the shade of trees, all the way eating fruits and drinking cold water...[13]

Religious institutions also contributed significantly in these efforts, by planting trees, digging wells, and other public facilities for all to use. There appeared to be no discrimination on any basis in the use of these facilities.[14] Ali Muhammad Khan has referred to the numerous mango orchards in the city of Ahmedabad which were maintained by the Muslims.[15] The rich people used to plant all sorts of fruits as well as ornamental trees in the courtyards of their houses.[16] *"A good house has its courtyard, gardens, trees..."*[17]

Both Terry and Nicholas have mentioned about the plantation of fruit trees by the rich people in their court yards and also in their groves. Terry writes that their places of burial was enclosed with a "*firm wall*". If possible they were constructed near some tank or spring of water. They contained "*pleasant*" fountains, little mosques, "*planted*" fruit trees and their noblemen's "choicest" flowers.[18] Father Monserrate writes that the city of Sirhind was beautified by many groves of trees.[19] A great care was taken to look after these trees.

Sultan Mahmud of Gujarat got the trees (planted in his garden) wrapped in red and green velvet and handsome women were engaged to look after them.[20] In the city of Farrukabad, there was a Naulakha bagh; a garden having reputedly 900,000 trees all within the walls. Apart from the gardens, trees in great numbers were planted inside the cities and in open spaces. Writing in 1803, Valentina remarks:

the trees most delightfully shade the houses and the open spaces.[21]

Flowers beds also occurred in some of those gardens primarily for their beauty and fragrance for example the gardens laid out by the aristocracy adjoining their mansions used to be so planned as to consist both fruit bearing trees and sweet scented attractive blossoms.[22] The Europeans in the Later period also took a great efforts to plant all sorts of trees especially the fruit trees such as guava, coconut trees, mango trees, etc.[23] Tree plantation was encouraged in all parts of the Empire. Abul Fazl writes that in Kashmir, people engaged themselves in planting fruit trees, and these were generally a source of benefit to the people.[24]

We get several evidences regarding the presence of the large tracts of forests in the Empire which were of great value in the progress of the general, but more so the urban economy of the Empire.[25] Most parts of India were covered by the forests growth which is the clear indication that forests and trees were not destroyed by the people. On the other hand they maintained them for fulfilling their needs.[26] The most stupendous and remarkable trees which were found in India were teak, the palm, the banyan, the *sisoo*, the *sal*, the *peepul*, the bamboo, the talipot, etc.[27] The sources of the Mughal period give a long list of the trees, flowers and fruits which were maintained. Babur, for instance, has given a long list of the trees, flowers and fruits which were found in India among these were the fruits such as mangoes, plantains, tamarind *(khurma-i-Hind)*, jackfruit, oranges, citron, etc.Trees such as *mahuwa*; whose wood was utilised for building purposes; date palm tree, *tar*, etc. and many flowers such as *jasun*, *kanir*, jasmine, etc.[28] Abul Fazl has mentioned about 72 kinds of wood which were available in India.[29] It shows that about 72 kinds of trees were available in India whose wood was utilized for various purposes, these were primarily *kanjak, khirni,* ebony, oak, *mahuwa,* sandal, *babul,* mulberry, *pipal, Bar,* etc.[30] He has also given a long list of the fruits and flowers which were available in the country. Fruits such as coconut, mangoes, grapes, melons, sugarcane and flowers such as *chambeli, ketki,* violet, *Gul-i-zafaran, kaner,* etc.[31] The large availability of these fruits and flowers indicate that these were maintained by the society on the large scale. Various ornamental trees such as cypress (*saru*), the pine (*sanubar*), the *chinar,* the white poplar and the *bid mulla* (willow) and sandal were grown on the large scale.[32] Mirza Haider Dughlat has stated that in Kashmir land was divided into four kinds and in one part; people laid gardens in which numerous fruit and decorative trees were planted and the other part was the "level ground, where the river bank abound in violets and many coloured flowers."[33] We also find a reference that during Aurangzeb's reign, Chowdary Mahesh, a wealthy person laid down a large garden near the Dal lake in Kashmir in which trees of all sorts were planted.[34] Edward Terry writes that the professionof the Parsis included all sorts of animal husbandry, sowing and selling of herbs, planting and husbanding vine and toddy trees, and looking after other fruit bearing trees.[35] De Laet has mentioned the extensive cultivation of palm trees at Broach, he writes:

It abounds in palm trees of forest variety...[36]

In Surat numerous gardens with fruit and ornamental trees were planted. In these gardens Indian merchants wandered on warm evenings *"to take air and feast in pleasant summer houses."*[37] A large number of orchards with numerous trees were planted throughout the country especially by the rich. Sujan Rai writes that in India there were so many gardens full of attractive trees.[38] Numerous inscriptions belonging to the period also testify the fact that both nobility and general public laid stress on the plantation of trees. For instance, there is an inscription on a well in the fort of Gulberga dated Monday 13th

December 1585, which states that mother of the Nawab Ibrahim Adil Shah, caused a well to be dug and an orchard to be planted at the shrine of Sayyid Muhammad Husaini, Gesu Daraz.[39] Also we get references from the various Mughal documents that a large number of trees were planted by the grantees in their milk land. One such document dated 25 Feburary1581 contains the information that a Parsi Mehr Tabib of *qasba* Navsari in sarkar Surat had planted about 109 *Khajur* (date) trees in his milk land. Another document dated 14th August1572 refers to the cultivation of 50 palm and 100 date trees.[40] Terry has written that in pleasure houses many pomegranate as well as all different types of flower and fruit trees were planted. In Sarkhej a numerous orchards containing fruit trees were planted by the wealthy people and the poor men were engaged to keep the orchards clean.[41] During this period the importance of the trees was realised on a large scale and stress was laid on the plantation of more and more trees.

Not only the trees were valued for the material gains even many trees were also regarded as sacred. In fact tree worship was possibly the earliest and the most prevalent form of religion in India. Tree worship remained a continuous religious practice throughout the historical period. Certain trees were scared, as they are in Hinduism today, notably the *pipal*, which was specially honoured by the Buddhists, under which Buddha found enlightenment. In fact Buddhism propagated love for trees. Its founder Gautam was born under an *asoka* tree, received enlightenment under the *Pipal* tree, preached his new gospel in mango groves, and under shady banyans and died in a *sal* grove. Also one of the seals of the Harappan period, depicts a horned Goddess on a *pipal* tree.[42] Lotus a symbol of purity was loved by Hindus. The cult of mother goddess which is pre-Aryan was adopted by Aryan Hindus, and the goddess Shri or Lakshmi, the spouse of Vishnu is associated with lotus. She is praised as "lotus born" (*Padma sambhava*), "lotus eyed" (*Padmakshi*), "lotus coloured" (*Padamvarna*), "standing on a Lotus" (*Padamsthita*), "decked with lotus garlands" (*Padamamalini*).[42a] During Mughal period also tree worship was practised on a large scale. Abul Fazl writes that *pipal* and *Bar* trees were worshipped by Hindus throughout the country.[43] He has narrated about the *Pipal* tree worship in Pattan Somnath. He writes:

Two and a half kos from Pattan Somnath is Bhal Ka Tirth(or the shrine of arrow). In this place an arrow struck Sri Kishnh and buried itself under a pipal tree on the banks of Sarsuti. This they call Pipal sri, and both these spots held in great veneration.[44]

Shade of the tree was regarded as one of the purifier that purifies the soul according to the Hindu *sahitya*.[45] Indians believed that the *pipal* was occupied by one or other of the Hindu triad, the gods of creation, preservation and destruction, who have the affairs of the universe to look after, but the cotton and other trees were occupied by some minor deities, who were vested with a local superintendence over the affairs of a district, or perhaps, of a single village.[46] Hindus also held the Banyan tree in special veneration, often assembled beneath its boughs, to perform ceremonies and sacrifices.[47] Della Valle writes about the banyan tree worship at Surat.

On an other side of the city... in an open place, is seen a great and fair tree, of that kind which I saw in the sea coats of Persia, near Ormuz, called there Lal but here Ber (Bar). The Gentiles of the country hold it in great veneration for its greatness and age, riting and honouring it often with their superstitious ceremonies...[48]

He further writes that the people had dedicated it to Goddess Parvati (mountain Goddess, wife of Siva). On its trunk, little above the ground a round circle was engraved, which *"had not any feature of a human countenance but according to their gross application represents that of their idol."* Its face was painted with *"bright flesh colour"*. Round about it flowers and pans were placed. These were frequently changed and replaced with fresh ones, the rejected stuff was bestowed as "sacred thing" to the people by the priest, who looked after the place, the people placed it upon their heads and kissed it.[49] A bell was also hung high up in the tree and was rung first by the devotees as they came. They also circled round the tree, some once, some twice, some thrice. When it was over they sprinkled rice, oil and milk before the idol.⁵⁰

Tavernier has also given a vivid account of the Banyan tree worship. He writes:"

*The Franks call it the tree of the Banians, because in place where there are any of these trees, the idolators sit under them and cook there food there. They reverence them specially, and generally build their pagodas either under or close to one of these great trees... and in its trunk which is hollow a monster is represented like the head of a deformed woman, whom they call Mamaniva.*⁵¹ *Every day a large number of idolaters assemble there to adore this monster, near whose shrine there is constantly some Brahman detailed for its service, and to receive offering of rice, millet and other grains made to it.*⁵²

Tavernier has explained the figures of deities represented under the banyan tree. He has marked them by numbers, for example number one is "the place where Brahmans dress up a representation of some one of their idol, as Mamaniva, Sita, and number two is the figure of Mamaniva..." and so on.⁵³ Sujan Rai has also given a detailed account of the religious importance of a *Bargad* tree which was grown in the fort of Allahabad.

*In the fort there is a tree. From the old days it is believed that this tree is holy and it is from the beginning and will remain till the end. It is called Akhi Bargad. Jahangir cut it down and put iron in it. However, it grew and surrounded the iron. Indians consider the temple there as most holy of the holy temples and during winter when the sun is in jadi they came here to dip. This is called Makra.*⁵⁴

There were many curious customs, traditions, and superstitions among the Hindus regarding trees. The marriage of the trees was a very common custom. Neither the man who planted a grove nor his wife could taste the fruit of a mango tree, until he had married one of the tree to some other, commonly the tamarind, that grew near it in the same grove. A great of pomp and show used to attend these unions; and of course, occasion was taken by the Brahminical priests.[55] Manucci narrates the importance of trees in the marriage ceremonies of the Hindus. He writes:

Then there is a certain tree in this country that they class as a Brahman, and style the marriage God. A branch of this tree is brought, and also one of other tree with bitter leaves, which they called Parechi (parisa),[56] meaning that it is the wife of other tree, according to them. The two branches are then interwoven round a post put in the centre of the arbour. There is also some sugarcane which they tie on. At the foot of all this are placed some leaves of a certain herb which is used to make powder.[57]

People had a great esteem for these trees especially the *pipal* that no Hindu dared and no Christian or Muslim condescended to lop off the heads of these trees.[58] Hindus had so much respect for these trees that they considered it improper to eat from the leaves of the *bar* or banyan and the *pipal* trees.[59]

Certain tree products were also regarded as sacred and were mainly used for the religious purposes. Flowers were the essential part of the morning worship. According to an aboriginal story-teller, flowers were not created for their beauty or scent or to adorn the beloved, they came into being so that man should have something to offer to Gods. This practice encouraged the growing of flowering trees like *champaka (michelia champaca)* and *Kadamba* in the precincts of temples. Most of the flowering trees which grow in India were sanctioned by the Aryans. The *Kadamba* was associated with Sri Krishna, *asoka* is dedicated to Kama Deva, the God of love, the red cup like flowers of the silk cotton tree were sacred to Shiva, the white flowers of *Kachnar* were associated with Lakshmi.[59a] Abul Fazl writes that for *Ishwar-Puja* or divine worship, Hindus mainly used sandal, flowers, betel, leaves, etc. They also fumigated the idols with perfumes.[60] Della Valle also refers to the offering of flowers and leaves to the God.⁶¹ From ancient times the fasts have been observed with religious fervour in Indian society. During fasts Hindus mainly took fruits. Abul Fazl has written that during different types of fasts, people mainly took water, fruits and milk.⁶² He writes about a kind of fast among the Hindus in which

*out of fifteen days, for three days and nights they eat only leaves, for three days and nights only the Indian fig; for three days and nights they are content with the seeds of the lotus; for three days and nights leaves of the pipal; for three days and nights; a kind of grass called drabha.*⁶³

Tree products also constituted an important part of the social occasions of every community. These products were widely used in the social ceremonies such as birth ceremonies, marriage, social parties, etc. Abul Fzal writes that the cradle of Akbar was made with Sandal wood and lign aloes and was decorated with rose leaves.⁶⁴ He writes that in the birth ceremony of Akbar

*… Fan wavers sprinkled otto of roses, and winnowed the air with sandal scented arms. Dark haired maidens freshened the floor by rubbing it with perfumes. Rose cheeked damsels gave a new lustre to joy by sprinkling rose water. Red garmented, sweet smiling nymphs enveloped the silver blossomed ones in gold by scattering saffron. Rose-scented, jasmine checked ones soothed the rapid dances with camphorated sandal wood.*⁶⁵

He writes that the trays of various coloured fruits were served among the nobles.⁶⁶ On their birthdays the Emperor used to distribute the fruits both fresh and dry among their courtiers.⁶⁷ Tree products were also frequently used in the Emperor's weighing ceremony (*tuladaan*). Mughal emperors celebrated their birth anniversaries twice in the year according to the solar and the lunar calendars and on these occasions they were weighed against numerous articles including the fruits which were later distributed among the poor. Abul Fazl describes the weighing of emperor Akbar on his solar birth ceremony:

His Majesty is weighed twelve times against the following articles: gold,... silk, perfumes..." He further writes *"His Majesty is weighed a second time on the 5th of Rajab, against eight articles, viz. silver... fruits...*[68]

In the *Tuzuk* there is reference to the solar weighing ceremony of Jahangir in which he was weighed against different tree products.

I was weighed twelve times, each time against a different item. The second weighing was against quick silver, the third against silk, the fourth against various perfumes like ambergris, musk, sandal wood, aloes, myrobalan, and so on for twelve weighing[69] writes Jahangir.

Fruits and flowers were frequently used both in Hindu as well as Muslim marriages. Manucci writes that when a marriage was fixed in the Hindus, the bridegroom's parents used to bring a bunch of flowers, a coconut and a branch of figs (banana) to be presented to the bride.[70] In the *Kshatriya* wedding the whole body of the bridegroom was thoroughly rubbed with powdered sandal wood, and was covered with flower necklaces sprinkled with leaves of gold.[71] The women used to anoint themselves with sandal wood unguent and decorated themselves with the garlands of flowers and pearls.[72] Flowers were frequently used in the Muslim marriages also. Pelsaert observes:

The bridegroom is dressed in red, and so garlanded with flowers and his face cannot be seen...[73]

He writes that the fruits and dry fruits were presented to the bride by her in laws:

Three or four days before it (marriage) the bridegroom and his parents go to the bride's house, with a great company of the whole tribe, and taking with them a large numbers of gonads, or large ornamented wooden dishes, full of confectionary, sugar, almonds, raisins and other fruits...[74]

Fruits were also distributed among the guests in the marriage ceremonies. For Instance, in the marriage ceremony of Prince Dara Shikoh:

on the 1ˢᵗ of sha'ban this year 1042 (11 Feb.1633), the ceremony of Hinabandi was performed… According to custom girdles of gold thread were distributed among those present and trays filled with conserves of roses, pan, argaja essence, and condiments and fruits of different kinds were brought into the assembly by the imperial domestics.[75]

Tree were also frequently used in the festivals. A festival very similar to *Holi*, called *Ab-i-Pashan* by Jahangir[76] or *id-i-gulabi* (rose water festival) by Abdul Hamid Lahori, was celebrated at the Mughal court with great elegance on the commencement of the rainy season. The princes and the prominent nobles would take part in the festival and greatly delighted in sprinkling rose water over each other. It was customary to present the king with jewelled golden flasks containing rose water, jujube tree flower juice and the aroma of the orange flowers on this festival.[77] Dry fruits were served frequently in the festivals. Inayat Khan writes about the serving of the dried fruits at Holy *Milad* festival:

By the order of the truth knowing Emperor, on the right of the 12ᵗʰ of Rabi/ this year 1043 (16 Sept.1633), an assembly was arranged in the forty pillared Hall of Audience, in pious observance of the auspicious milad, or birthday ceremony of the Prophet. Various scholars and pious persons recited the Quran… and for enjoyment of the assembled worthies, the atmosphere was perfumed by incense and fragrant essences, and they were served banquet trays of varied foods, dry fruits and sweets[78]

During the reign of Akbar Shah II (1806-1837), a festival called *Phool walon ki sair* (festival of the flower sellers) started at Delhi, and soon it became extremely popular with all citizens, from artisans to aristocrats, who would without exaggeration, all leave their work and join the celebrations and during the reign of Bahadurshah Zafar

(1837-1857), the last Mughal Emperor, the festival reached its peak in the participation of the court and people of all communities. On the first day of the festival, there used to be a procession of fans and flowers.[79] Fruits were sold in the religious fairs. Sujan Rai Bhandari while describing a fair at Batala writes:

Two kos from Batala is Achal, a place sacred to syam kartik, it is an old shrine... thousands of mendicants of austere devotion and many anchorites bent up benefiting others, come to this place.. and in this merry gathering, in one part of the bazaar, on the two sides of the road, are arranged on trays and dishes many kinds of eatables, confects, fruits of spring and autumn, perfectly sweet and fresh.[80]

Fruits and flowers were frequently used in the worship of God. For instance the worship of Mahadeva figures prominently in the *Basant Panchami* festival. The young maidens used to offer fruits and flowers to the temple of Siva, where the emblem of Siva was washed with sandal and aloe-wood paste.[81] We also get references to the decoration of the fruits in the shops of Shahjahanabad for the welcome of Aurangzeb on his return to Shahajahanabad. Various kinds of fruits, fresh and dry were arranged very nicely in the shops of Shahjahanabad.[82] In the social parties or *jashans* tree products were used on a very large scale. Aloe-wood and incense were constantly burned there. Rose water was frequently sprinkled over the party for its refreshing and cooling effects. Fruits both fresh and dry were served in silver and golden fruits trays.[83] On arrival a guest in the house of a Hindu was welcomed with special forms. In ordinary cases betel leaf and flowers were offered to visitor. In the case of the distinguished visitor a platform was raised, flowers were strewn over it, and sandal wood paste was held in readiness to rub on forehead.[84] Describing about a feast organised by Asaf Khan, Terry writes that fruits formed the main part of the feast,

"Large dishes were composed of... many salads of curious fruits, some preserved in sugar, other raw.[85]

Fruits formed the important part of the feasts. Describing about the lavishness of one such feast organised by Nurjahan, Jahangir writes:

... on Thursday the 26th... I held a celebration in one of the pavilions occupied by Nurjahan Begum in the middle of a large lake. The amirs and intimates were invited... All sorts of roasted meats and fruits were placed before everyone as relishes.[86]

At a splendid banquet given in the honour of Akbar by Asaf Jah and Muzaffar Nishan a large varieties of delicious dry fruits were served.[87] Thomas Roe gives an account of the dinner to which he was invited by a noble Mir Jamaluddin Hasan. He writes that the dinner composed of

dishes of diverse sorts, raisins, almonds, pistachios and fruits.[88]

It was customary to gift tree products to the friends, relations and dignitaries on the different occasions especially by the upper class. The Mughal sources records several evidences pertaining to tree products as gift. Bheeram (Ghoggar) Dev, the ruler of Jammu used to sent yearly to the Court of Delhi, for Babur, a well stocked bag of wild walnuts, onions, etc. as a token of friendship.[89] Jahangir was presented by seven camels of rose water by the Shah Abbas of Iran.[90] It was also usual that nobles used to sent the gifts of tree products to the emperors as the token of loyalty. Afzal Khan of Bihar sent to the court of Jahangir, sandal wood, musk bags. Aloe woods, along with many other gifts.[91] There is a reference in the memoirs of Jahangir that Rana of Mandu presented to Jahangir several jars of pickles and marmalades along with other items.[92] Sir Thomas Roe mentions about the gifts of fruits such as melons, pines, plantains, plums, etc. by Jahangir to the Persian ambassador[93] Aurangzeb used to send *Dallies* (baskets) of grapes to Shahjahan as a token of loyalty from Deccan.[94] There are numerous references form the reign of Aurangzeb which show that fruits and perfumes were sent as gifts by the rulers of other countries to the court of Aurangzeb as a token of friendship. About the gifts of the King of Persia to Aurangzeb, Manucci writes:

This present consisted of…, sixty cases of perfect rose water, and 20 cases of another waster distilled from a flower which is only found in Persia, and is called bedemus (bed-i-mushk) (Egyptian willow); it is a very comforting water against all fever caused by heat…[95]

In addition to this Shah of Persia sent perfumes, 400 trays of sweets and fruits, Persian melons (*Kharbuza-i-karez*) and varieties of fresh and dry fruits[96] Another reference from Aurangzeb's reign shows that king of Balkh sent his ambassador to the court of Aurangzeb along with gifts fruits to establish a "firm friendship."[97] About this Manucci writes:

… the third present consisted in 100 camels loaded with fresh fruits – melons, apples, pears, pomegranates, and grapes without seeds; and other 100 camels loaded with dried fruits – Bukhara plums, the best in the world, apricots, quismis (kishmish or raisins), which are white, seedless grapes of great sweetness, and three kinds of dried grapes, one large an white, which looks candied and the Italians called zebibo, and the other two kinds purple – one large, the other small both very sweet and nuts, filberts, pine nuts, almonds and pistachios. Aurangzeb showed pleasure at these presents…[98]

Sujan Rai writes that Nishkar or sugar cane was also given far and wide as a good gift.[99] Jahangir writes of receiving a present of 1500 melons from the Khan Alam of Kariz;

On this day there arrived 1,500 melons from kariz. The Khan Alam had sent them as a present. I gave a thousand of them to the servants and five hundred to the women of the harem.[100]

Fruits were also given as rewards for the services. Manucci was given the reward of fruits for treating a sick man. He writes:

Although, against my will, I went on with my treatment of sick man'.... Continuing with some tonic extract of coral, I restored him to health in 5 days and the envoy was so pleased that he made me a present of nine melons and a quantity of dried fruits[101]

On his victory against Afghans, Maharaja Jaswant Singh was rewarded by Aurangzeb who conferred upon the Maharaja a special robe *(khilat-i-khasa)* with an elephant worth rupees 20,000, sword and seven trays *(khwan)* of fruits.[102] Tavernier was also presented a gift of fruits by the king of Bargant.[103] He writes:

On one of my journey to Agra, when passing by Bargant, I did not see the Raja, but only his lieutenant, who treated me with great civility, and presented me with rice, butter and fruits of season.[104]

Fruits and fruit products were also given as charity; Asad Khan, an Afghan noble used to give a variety of pickles, relishes and betel-leaves as charity to the beggars.[105] There is a reference that Aurangzeb when he was in Deccan regularly sent mangoes to Shajahan from Burhanpur as gifts.[106]

Tree products were not only utilised on the social occasions but they also formed an important part of the day to day life of the people. People were very much dependent on forests for their needs. The forests which were within the reach of the towns or villages furnished the inhabitants with timber, fuel and minor produce. Since there were more forests and less cultivation, the larger portion of the rural population was very much dependent on the forest produce.[107] These products were used by the people on the large scale. Bernier was amazed to see the great composition of both fresh and dry fruits at Delhi.[108] Father Monserrate writes that fruits constituted the main part of the food of the Parsis of Surat.[109] The seasonal fruits such as mangoes, black berries, oranges, cucumber, guava, dates, figs, grapes, etc. were enjoyed by rich and poor alike.[110] Certain fruits were also imported from the foreign countries to meet the needs. But being very dear they were used only by the rich. According to Bernier

Nothing is considered so great a treat, it (fruits) forms the chief expanse of the Omrahs

and he goes on to cite the instance of his *Agah* (master) who would not mind spending 20 crowns for his breakfast alone.[111] Apart from fruits different types of pickles such as mango achar, plantain soups, etc. were taken largely by the people.[112] For toilet purposes sweet scented oils made from different flowers, sandal wood paste mixed with saffron and many scents and rose water were used by the people.[113] Perfumes like *Santak* (a product of civet, chuwah, chambeli essence and rose water) and *argajah* were used on a large scale. People used a sweat powder made of sandal wood to get the sweat out of their bodies and head.[114] In the *hamamas* (baths), oils, perfumes, essences of sandal, cloves and oranges were freely applied to the customers. The people kept their feet as clean and soft as their hands by anointing them with scented oils.[115] Pelsaert writes that in summers sandal wood and rose water was frequently used for bathing purposes.[116]

Tree products were so much in demand that there are some incidences of the plundering of these products by the people. During the reign of Babur we come across the incidences of plundering of the fruits coming from Kashmir to the court of Delhi. The Kashmir fruit *dallies* (loads or baskets) were being more than once plundered on their way to the court of Delhi by some of the numerous predatory clans and tribes then infesting and inhabiting the different parts of the route, such as Jellall tribe, at or about Rajouri, the Chib tribe at Bhimber, etc.[117] Another incidence of the plundering of sandal, lignum aloes and bagam wood, etc. by the royal army at Bijapur has been recorded by Inayat Khan.[118] Many instances show that people were dependent on trees for their livelihood. Many fruits that grew in the jungles were gathered by the poor for their sustenance.[119] There were Jungles of mangoes, khrini and tamarind while entering Gujarat via Dohad and *"faire woods of Kheernee, peelooes, etts."*, as well as *"mangooes"*, when entering the province from the direction of Sirohi writes Peter Mundy.[120] "wild date trees" grew between Baroch and Surat.[121] Barbosa who writes in the beginning of the 16th century tells us that the people at Malabar coast were very poor and were dependant on forest produce for their livelihood; some bringing wood and grass for sale in the city, others living on roots and wild fruits, covering themselves with leaves.[122] Many fruits notably the melon *(kharbuza)* were cultivated by the peasants, who used to sell them in the markets. In Deccan helpless and destitute people used to grow muskmelons *(kharbuza-i-garma)* in the sand upon the banks of the river.[123] Trees bearing the better class of fruits such as quality mangoes were usually planted in groves. The groves might belong to the peasants, but it is probable that quite often they were owned by the rich people, who seasonally rented them out to the cultivators and professional fruit sellers. This was the case in Goa, where the Portuguese *"let out (their coconut trees) unto the canariins"* some of these rentiers having as many as 300 or 400 trees or more.[124] Members of aristocracy and officials possessed orchards to have fruit not only for their own consumption but also to sell for profit.[125] The saffron crop of Kashmir has for centuries been important economically, while the gardens of Dal lake in kashmir in the days of Jahangir produced a substantial income from roses and bed musk.[125a] Even the great imperial garden at Sirhind was yearly rented out for fifty thousands *rupees*.[126] Muslims, built their graves amidst groves of the fruit trees, the income from which went to support their descendants or the guardians of their groves.[127] Generally the control of the grave yards was in the hands of *madad-i-maash* holders and income from the fruits was appropriated by them. Thereby with the plantation of orchards *madad-i-maash* holders improved their economic conditions and also provided fruit to the locality.[128] Groves played an important part in the rural economy of some districts where they accounted for as much as 5% of the total cultivated acreage and provided a significant source of income for the lands holders.[129] Tree products, therefore were the source of livelihood of the people and also helped to strengthen the economy of the empire.

REFERENCES AND NOTES

1. Fergusson, James. *Tree and Serpent Worship*. 2nd edition London: India Museum, W.H.Allen and co. publishers, 1873 p-2
1a. Randhawa. *Gardens Through the Ages*, p-46.
2. *Ibid*, p-48.
3. *Ibid*, pp-53, 54.
4. Naqvi. *Urbanisation and Urban Centres*, pp-41, 42.
5. Irfan Habib. *Agrarian System*, pp-53, 54.
6. Mohammad, Jigar. *Revenue Free Land Grants in Mughal India Awadh Region*. New Delhi: Manohar publishers, 2002 p-88. Hereafter Jigar Mohammad.
7. C.A.Bayley, p-43.
7a. Mohammad, Jigar. *A Panoramic Reconstruction of Sufism in the Jammu Hills, Sufism in Punjab Mystics, literature and Shrines*. Edited by Surinder Singh and Ishwar Dayal Gaur. New Delhi: Aakar, 2009 pp-123-129.
7b. *Ain* vol-3, p-355
7c. *Tuzuk-i-Jahangiri* vol-II pp-149,150
8. K.M.Ashraf. *Life and Conditions*, p-206.
9. *India under British Empire*. p-93.
10. William Foster. *Early Travels*, p-14.
11. Pelsaert. *Jahangir's India;* p-48.
12. Bernier, p-399.
13. *Khulasat* in Geographical Glimpses, Vol.3, p-61.
14. *Geographical Glimpses*, Vol.3, intr.xxiv.
15. Geographical Glimpses, Vol.3, *Mirat-i-Ahamdi* p-45.
16. Father Monserrate, p-219.
17. Bernier, p-247.
18. William Foster. *Early Travels*, pp-289, 290.
19. Father Monserrate, p-102.
20. Nizamuddin. *Tabqat-i-Akbari*, p-393.
21. Muhammad Umar. *Urban Culture*, pp-24, 25.
22. Naqvi. *Urbanisation and Urban Centres*, pp-43, 44.
23. *British Social Life in India*, p-62.
24. *Ain*, Vol.2, p-355.
25. Naqvi. *Urbanisation and Urban Centres;* p-45.
26. Moreland, W.H. *India at the Death of Akbar: An Economic Study*. New Delhi: Atlantic Publishers and Distributors, 1994 p-135.
27. *India Under British Empire*, p-93.
28. *Baburnama*, pp-503-515.
29. *Ain*, Vol.1, pp-237-239.
30. *Ibid*, p-

31. *Ibid*, pp-68-83.
32. *Tuzuk*, Vol.1, pp-6-7.
33. *The Tarikh-i-Rashidi*, p-424.
34. *Aurangzeb in Kashmir*, p-4.
35. Ansari. *European Travelers*, pp-98-99.
36. De Laet, p-24.
37. *British Social Life in India*, p-11.
38. *Khulasat* in Geographical Glimpses, Vol.3, p-61.
39. Bendrey. *Muslim Inscriptions*, p-133.
40. Tirmizi. *Mughal Documents*, pp-59, 56, 52.
41. Ansari. *European Travelers*, p-86.
42. Basham. *The Wonder that was India*, p-23.
42a. Randhawa. *Gardens through the Ages* pp-48,49
43. *Ain*, Vol.1; p-92.
44. *Ain* Vol.2, p-253.
45. *Ain*, Vol.2, p-320.
46. *India Under British Empire*, p-96.
47. *Ibid*, p-94.
48. Della Valle Vol.1, p-35.
49. *Ibid*, pp-36, 37.
50. *Ibid*, p-37, 38.
51. Tavernier, Jean Baptiste. *Travels in India* Vols.I, II and III. Trans. by V.Ball. New Delhi: Orient Books Corporation,1977. Vol.II p-155, f.n. 2; The Term probably Manamai or Mama Devi, the mother of Gods.
52. *Ibid*, pp-154, 155.
53. *Ibid*, pp-155-156.
54. *Khulasat* in Geographical Glimpses, Vol.3, p-71.
55. *India under British Empire*, p-99.
56. Thespesia Populnea, the flowering pipal, palas pipal, or tulip tree; Manucci Vol.3, p-59.
57. Manucci Vol.3, p-59.
58. *India under British Empire*, p-95.
59. *Ain* Vol.3, p-325.
60. *Ibid*, pp-301, 302.
61. Della Valle Vol.1, p-37.
62. *Ain* Vol.3, p-326.
63. *Ibid*, p-327.
64. The *Akbarnama*, Vol.1, p-132.
65. *Ibid*, p-57.
66. *Ibid*, p-58.
67. Pran Nath Chopra, *Some aspects of Society and Culture*; p-92.

68. *Ain* Vol.1, p-277.
69. The *Jahangirnama*, p-197.
70. Manucci Vol.3, p-53.
71. *Ibid*, p-59.
72. *Ain* Vol.3, p-342.
73. Pelsaert. *Jahangir's India*, p-82.
74. *Ibid*, p-82.
75. The *Shahjahan-nama*, p-91.
76. The *Jahangirnama*, p-221. A very illustrate painting No.636, I.A.E., Mughal; Early 17th century shows, "The Emperor Jahangir celebrating the festival of *Aab-Pashi* or sprinkling of rose water", painted by Govardhan on the 5th Amardad day. Flasks full of rose water, white, yellow and blue are before Jahangir while around him stand courtiers and some ladies.
77. Pran Nath Chopra. *Some aspects Society and Culture*, p-95.
78. The *Shahjahannama*, p-118.
79. Maheshwar Dayal. *Rediscovering Delhi*, pp-200-201.
80. *Khulasat.* p-92.
81. K.M.Ashraf. *Life and Conditions*, p-238.
82. *Futuhat-i-Alamgiri*, p-114.
83. K.M.Ashraf. *Life and Conditions*, p-229.
84. *Ibid*, p-270.
85. Ansari. *European Travelers*, p-88.
86. The *Jahangirnama*, p-224.
87. Mohammad Arif Qandhari. *Tarikh-i-Akbari*, p-177.
88. Ansari. *European Travelers*, pp-73, 74.
89. Symth, G.C. *A History of the Reigning Family of Lahore.* Delhi: Parampara Publications,1979. p-239.
90. *The Jahangirnama*, pp-121, 122.
91. *Ibid*, p-127.
92. *Ibid*, p-221.
93. Foster, William ed. *The Embassy of Sir Thomes Roe to India*(1615-1619). As narrated in his journal and correspondence. New Delhi: Munshiram Manohar Lal Publishers, 1990 p-364.
94. Razi, Aqil Khan. *Waqiat-i-Alamgiri*. Delhi: Mercantile Print, Press, 1946 p-13.
95. Manucci Vol.2, p-46.
96. *Aurangzeb in Muntakhab-ul lubab*, p-176.
97. Manucci Vol.2, p-32.
98. *Ibid*, p-34.
99. *Khulasat* in *Geographical Glimpses*, Vol.3, p-62.
100. *Tuzuk* Vol.1, p-435.

101. Manucci Vol.3, p-37.
102. *Futuhat-i-Alamgiri*, p-111.
103. Tavernier Vol.2, p-31. F.n.3; Bargant called Bergam; the proper name is probably Baglana.
104. *Ibid*, p-69.
105. K.M.Ashraf. *Life and Conditions*, pp-258-259.
106. Faruki, *Aurangzeb and His Times*; p-16.
107. *India at the Death of Akbar*, p-135.
108. Bernier, pp-203, 204.
109. Father Monserrate, p-8.
110. P.N.Chopra. *Some aspects of social life during the Mughal Age (1526 – 1707)*, p-36.
111. Bernier, p-249.
112. Ojha, P.N. *North Indian Social Life during Mughal Period*. Delhi: Oriental Publishers and Distributors, 1975 p-7. Also see *British Social Life in India*; p-17-18.
113. P.N.Chopra. *Some Aspects of Social Life*, pp-17, 18.
114. *Ibid*, p-18.
115. *Ibid*, p-18.
116. *Jahangir's India*, p-65.
117. *A History of the Reigning Family of Lahore*, p-237.
118. *The Shahjahan-nama*, p-134.
119. Irfan Habib. *The Agrarian System*, p-53.
120. *Ibid*, p-53, f.n. 105.
121. *Early Travels*, p-175.
122. *India at the Dealth of Akbar*; p-250.
123. Irfan Habib. *The Agrarian System*, pp-53, 54.
124. *Ibid*, p-54.
125. *Ibid*, p-54.
125a. Sylvia Crowe, p-193
126. *Early Travels*, p-158.
127. Irfan Habib. *The Agrarian System*, p-54.
128. Jigar Mohammad. *Revenue Land Grants*, p-88.
129. C.A.Bayley, p-43.

BOTANICAL PRODUCTS AND THEIR SIGNIFICANCE

Mughals fully realised the importance of the forests and trees in the general economy and prosperity of the country and followed a definite policy for their conservation and scientific exploitation. From the 16th century onwards, the trees were more properly utilised than the previous centuries. Trees were associated with all important socio-economic activities of Medieval India. It is evident that most of the contemporary sources of the Mughal period contain description of trees and tree products.[1] The trees contributed to the development of both the rural and urban economy of the empire. The trees not only provided cheap fuel, but made available timber, fruits, bamboos, fibres, gums, sandal and other fragrant woods, drugs, etc. As the growth in urbanisation gathered momentum, the construction work specially of the durable buildings had grown enormously. Therefore there was an increase in the consumption of timber in the buildings of empire.[2] The botanical products were related with almost every important activity of the 16th and 17th century India. The manufacture of the means of the transport, building materials, maintenance of the silk industry and manufacture of various other articles were dependent on the trees. Since during the 16th and 17th centuries the demand of all these items increased the utility of the tree products also increased.

The contemporary sources of the 16th and 17th centuries mention that the trees yielded numerous products for various uses. The most important product of the trees was the fruits. Contemporary sources have given a long list of the fruits grown in India. Babur has enumerated a detailed account of the fruits grown in India among these were mangoes, plantain, *anbali*, *mahuwa*, mimusops, *jamans*, *jack*-fruit *(kadhil)*, monkey jack, lote fruit, *karunda, amla,* date-palm, coconut-palm, *tar,* oranges, lime, citron, *sangtara, sadafal,* lemon, etc.³ The *Ain-i-Akbari* has also enumerated about sixty types of fruits cultivated in India such as mulberries, gular, pineapples, oranges, sugarcane, *bers, usiras, bholsairs,* mangoes, jack-fruit, figs, melons, *mahuwa,* plantains, dates, pomegranates, guavas, water-melon, grapes, etc.⁴ The prices of fruits were very cheap for example, the price of 100 mangoes was 40 *dams* (i.e one *rupee*); 80 to 100 *dams* for one *man* pomegranates (*anar*) and 4 *dams* for one *seer* dates. ⁵ These prices indicate a large availability of fruits. A long list of *Mewa-i-Turan* (Turani fruits) is also given by Abul Fazl. These were very popular among the noble class especially Kabul melons, cherries *(shahalu),* apples, seed less pomegranates, apples, pears, quinces, guavas, peaches, apricots, *aluchas,* almonds, plums, figs, raisins *(kishmish), Jawz* (nuts), plums, etc⁶ William Finch has mentioned the cultivation of the fruits like apples, *toot* (*shahtoot,* Mulberry), almonds, peaches, figs, grapes, quinces, oranges, lemons, pomegranates, etc.⁷ in the royal gardens. From the account of Peter Mundy (1632) it is evident that there was a large availability of fruits in India such as apples, oranges, mulberries, mango, figs, plantain, cherry, apricot, pomegranates, guava, etc.⁸ A long list of *Mewa-i-Hind* or Indian fruits has been given in the *Ain,* these were mangoes, pineapples, oranges, sugarcane, jackfruit, plantains, *ber,* guava, figs, mulberries, custard-apple, water-melons, *mahuwa, tendu, gumbhi,* dates, *gal-gal, jaman, phalsa, singhara,* carrot, *garnal, kanku, ambili, salak, peth, senb, tura,* etc. All these fruits were cheap and were consumed by the larger section of the society.⁹ Kashmir and Kabul produced fruits of the colder climates, while the rest of the empire was better endowed with the fruits of warmer regions such as bananas, oranges, mangoes and guavas, etc.¹⁰ Among the fruits mentioned by Jahangir which were available in India were melons, mangoes, grapes, - *sahibi, habshi* and *kishmishi* were common in several towns and pineapples were also available.¹¹ Bernier has given a long list of the fruits that were available in the fruit market of Delhi. In Summers the shops were well supplied with dry fruits from Persia, Balkh, Bokhara and

Samarqand, such as almonds, pistachios, walnuts, raisins, apricots, etc. and in winter there were excellent fresh grapes – black and white, pears and apples of three or four sorts and admirable melons which lasted the whole winter.[12] In all parts of the Empire a wide variety of fruits and flowers were grown. According to Sujan Rai:

If one describes the multi-coloured and well tasting fruits of the land. It will comprise of a separate book.[13]

Tavernier too has referred to a wide variety of the fruits grown in the regions like Kolaras, Ahmedabad and Goa, such as mangoes, *ananas*, figs, plantains and coconuts.[14] A variety was added in the list of the fruits with the introduction of new fruits such as pineapple,[15] guava and custard apple,[16] papaya and cashew-nuts,[17] sweet cherry,[18] etc. In later period also India produced all these fruits such as mangoes, pineapples, plantains, pomegranates, pumplenoses, jacks, custard-apples, leeches, guavas, melons, oranges, lemons, limes, grapes, soursops, almonds, goose-berries, strawberries, tamarinds, plums, figs, dates, citrons, loquats, mangoes, most of these fruits were found on the tables of the Europeans who settled in India.[19]

Wood was another important tree product. The trees provided large quantities of the wood which was used as timber and fuel. Though Pelsaert writes that:

Fire wood is consequently very dear and is sold by weight, 60 lb, for from 12 to 18 piece (or 5 strives), making a serious annual expanse for a large house hold. [20]

But Pelsaert seems to have highly exaggerated and inflated the prices because fire wood was largely available in the country. From the account of Abul Fazl one finds that 600 wagons were used in the space of 10 months to carry fire wood to the royal kitchen. The quantity of the consumption of fire wood in the royal kitchen was 150,000 *mans*.[21] It appears that Pelsaert's observation is connected with those towns where supply of the fire wood was scarce because of the transport difficulties. However, trees' yield in form of timber was great. Babur has described *mahuwa* wood as most popular for house construction.[22] *Ain* mentions seventy–two types of trees providing timber. *Kanjak* wood has been mentioned to be the heaviest and the *safidar* as the lightest wood.[23] The weight of the one cubic *gaz* of Kanzak was 27 *mans* and 14 *sers* and of *safidar* 6 *mans* 7 *sers* and 22 ½ *tanks*.[24] Although the peepal tree is known as the sacred tree of India, its wood was also used for building purposes.[25] The 72 types of wood mentioned by Abul Fazl along with their weights are:[26]

S.No.	Name of the Wood	Weight		
		Mans	*Sers.*	*Tanks*
1.	*Kanzak*	27	14	-
2.	*Ambli (tamarindus indica)*	24	8 ¾	25
3.	*Zaytun (Gyrocarpus asiaticus)*	21	24	-
4.	*Balut (Oak)*	21	24	-
5.	*Kher (Acacia catechu)*	21	16	-
6.	*Khirni (Mimusops)*	21	16	-
7.	*Parsiddh*	20	14	17
8.	*Abnus (Ebony)*	20	9	20
9.	*Sain (Acacia Suma)*	19	32	10
10.	*Baqam (Caesalpina sappan)*	19	22 ½	10
11.	*Kharhar*	19	11 ¼	5
12.	*Mahwa (Bassia Latifolia)*	18	32 ½	2
13.	*Chandani*	18	20 ½	10
14.	*Phulahi*	18	20 ½	10
15.	*Red Sandal (Rakt Chandan)*	18	4 ½	10
16.	*Chamri*	18	2	7 ½
17.	*Chamar Mamri*	17	16 ¼	-
18.	*Unnab (Zizyphus Sativas)*	17	5	4
19.	*Sisau Patang*	17	1 ¾	7
20.	*Sandan*	17	1	28
21.	*Shamshad (Buxus sempervirens)*	16	18	25

22.	Dhau (grislea tomentosa)	16	1	10
23.	Amla (Emblica officinalis)	16	1 ½	1
24.	Karil (sterculia fetida)	16	1	10
25.	Sandal	15	17	20
26.	Sal (Shorea robusta)	15	4 ¾	7
27.	Banaus (Cherry, Shah Alu)	14	36 ½	10
28.	Kailas (Cherry – tree)	14	35 ½	-
29.	Nimb (Azadirakhta indica)	14	32 ¼	31
30.	Darhard (Berberis aristata)	14	32 ¼	19
31.	Main	14	22 ¾	-
32.	Babul (Acacia Arabica)	14	22 ¾	-
33.	Sagaun	14	10	20
34.	Bijaysar	13	34	-
35.	Pilu	13	34	-
36.	Mulberry	13	28 ½	15
37.	Dhaman	13	25	20
38.	Ban Baras	13	10	29
39.	Sirs (Acacia odortissima)	12	38	21
40.	Sisau (Dalbergia sissoo)	12	34 ¼	5
41.	Finduq	12	26	4
42.	Chhaukar	12	17 ½	22
43.	Duddhi	12	17 ½	22
44.	Haldi	12	13 ½	32
45.	Kaim (Nauclea parvifolra)	12	12 ½	30
46.	Jaman (Jambosa)	12	8	20
47.	Faras	12	8	20
48.	Bar (Ficus indica)	12	3 ¼	5
49.	Khandu	11	29	-
50.	Chanar	11	29	-
51.	Charmaghz (Walnut tree)	11	9 ¼	17
52.	Champa (Michelia champaca)	11	9 ¼	17
53.	Ber (Zizyphus jujuba)	11	4	-

54.	Amb (Mango)	11	2	20
55.	Papari (Lumus)	11	2	20
56.	Diyar (Cedrus deodar)	10	20	-
57.	Bed (Willows)	10	20	-
58.	Kunbhir (gmelina arbora)	10	19 ½	22
59.	Chidh (Pinus longifolia)	10	19 ½	22
60.	Pipal	10	10 ¼	21
61.	Kathal (Jack Tree)	10	7 ½	34
62.	Gurdain	10	7 ½	34
63.	Ruhera (Terminalia belerica)	10	7	30
64.	Palas (Butea frondosa)	9	34	-
65.	Surkh Bed	8	25	20
66.	Ak (calotropic gigantean)	8	19 ¼	25
67.	Senbal (Cotton tree)	8	13	34
68.	Bakayin (Melea composite)	8	9	30
69.	Lahsora (cordia mixa)	8	9	20
70.	Padmakh (Cerasus caproniana)	8	9	20
71.	And	7	7	31
72.	Safidar	6	7	22 ½

In above weights the *ser* has been taken as 28 *dams*.

Bamboo was the another tree product which was mainly used for covering the ceilings.[27] Bamboo again was available in different qualities. First quality, 15 *dams* for 20 pieces; 2nd quality, 12 *dams* for do; third quality, 10 *dams* for do.[28] The price of some kinds of Bamboo was much higher. A peculiar kind was sold at 8 *Ashrafis* (*muhrs*) per piece.[29] There were different qualities of *Patal* which were made of reed. 1st quality, cleaned 1 ½ *dams* per square *gaz*; 2nd quality, 1 *dam* per square *gaz*. Some times *patal* was sold at 2 *dams* for piece 2 *gaz* long and 1 ½ *gaz* broad. *Sikri* was made of very fine *qalam* reeds were sold at a rate of 1 ½ d. per pair, 1 ½ g. long and 16 *girihs* broad.[30]

Jahangir has given a vivid account of the trees whose wood was used for various purposes. Those were:

... the cypress (sarw), the pine (sanubar), the chanar (plantanus orientalis), the white poplar(safidar, populus alba), and the bid mulla (willow)... The sandal tree which once was peculiar to the islands (i.e. Java and Sumatra, etc.), also flourishes in the gardens. [31]

Different parts of the empire yielded different types of the timber. For instance, from the forests around Kalinjar black coloured ebony was obtained. Good quality timber was also available in the sarkar of Bazuha (in Bengal), in Gujarat forests and also in *subah* of Lahore.[32] In Bengal and Orissa several varieties of Bamboo which was used for different purposes was available.[33] Trees like *sissoo, teak* or *sygwan* and *tal-ipot* or *talpat* afforded the timber which was put to various uses.[34] Different types of the fragrant woods such as sandal wood and aloes were grown in Hindustan. Assam was specially reputed for a peculiar quality of aloe-wood which was sent as an offering to some of the most famous temples in the land.[35]

Another important botanical product were the flowers. These were utilised for the manufacture of different types of the scented oils and perfumes. Babur has greatly praised the variety of flowers available in Hindustan such as *jasun* (hibiscus rosa sinensis), *kanir* (oleander), *kiura* (the screw pine), jasmine, etc[36] Abul Fazl has given a long list of the flowers grown in India. He has put them in different categories – First category includes the fine smelling flowers, e.g. *sewti, chambeli, raybel, mongra,* the *champa,* the *nargis,* the *kewara,* the violet, etc.[37] Second included the flowers notable for their beauty such as *gul-i-Aftab,* the *gul-i-kawal,* the *kesu,* the *senbel,* the *ratanmala,* the *nag-kesar,* the *champala,* the *lahi* etc [38] and the third category included various Irani and Turani flowers introduced in India such as *gul-i-surkh,* the *nargis,* the *yasman-i-Kabud,* the *rayhan, etc.*[39] Jahangir is also full of praise for the flowers of India and has remarked that no other flowers of the world can be compared with those of India.[40] According to him very fine flowers were grown in India especially in Kashmir such as red roses, violets, narcissi, etc.[41] William Finch has mentioned the cultivation of the flowers like roses, stock-yellow-flower, marigolds, wall-flowers, ireos, pinks, white and red, etc.[42] All the flower found in India were of very attractive and diverse colours with fine and refreshing aroma. These flowers were used on all happy occasions like weddings, births, feasts, festivals, etc.[43] Besides these many other flowers such as roses, water-lily, violet, Egyptian willow, narcissus, jasmine, etc. were also found.[44]

In addition to above mentioned products there were large number of other botanical products which were obtained from the trees such as cloves, spikenard, sandal, amber, camphor, gumlac, bees wax, saffron, etc.[45] Gumlac was a Kind of wax found on the barks of certain shading trees.[46] Bernier has described the lac of Bengal was of best quality. He also describes the wax of Bengal as the best.[47] Tavernier describes that gum lac was one of the important trade items at the port of Surat. He has described about the two types of the gumlac – one washed, which was per *maund* for 10 *mahmudis* and the other was in sticks of sealing wax per *maund* for 40 *mahmudis*.[48] Gumlac was also received from the forests of Gujarat, Malabar, Gingeli,[49] Bihar, Orissa, Assam, Malwa and Bijapur.[50] Abul Fazl has mentioned about various types of the gums which were obtained from the trees, one kind was *Silaras* (Storax). In Arabic it was called *Misah*. It was the gum of a tree that grew in Turkey. There were two types of this wax. The Kind which was clear was called *Misah-yi-sayila* (liquid); the other kind *Misah-yi-yabisa* (dry). The best kind was that which spontaneously flowed out of the trunk; it was yellowish.[51] Other type was *Luban* (frankincense). It was the odorous gum of a tree found in Java.[52]

Camphor was the another product which was available from trees. The camphor trees were, generally, found in the *ghauts* in Hindustan and in China. Camphor was collected from the trunks and branches. The camphor within the trees looked like a small bits of salt; that on the outside like resin. It often flowed from the trees on the ground; and gets after sometimes solid. If there were earthquakes during the year or any other cosmical disturbances, camphor was found in large quantities. Of various kinds of camphor the best was called *Rihabi* or *qaysuri,* inferior to it was the kind called *Qurqury,* which was blackish and dirty.[53] Sandal which was called *Chandan* in Hindi, was also widely used across the country especially for the religious purpose. The tree was mostly found in China. But during Mughal period, it was successfully planted in India. There were three kinds – the white, the yellow and the red. Some took red to be more refreshing than the white, other preferred the white. The latter was certainly cooling than the red, and the red more so than the yellow. The best was that which was yellow and oily; it goes by the name of *Magasari.*[54]

During the medieval period the significance of the tree products was realised to a greater extent. The tree products were of great value in all spheres of man's life. Trees provided shelter against inclement weather, gave fruits and nuts to eat, and wood for implements and building. Appreciating the value of trees Abul Fazl writes:

It would be impossible to give an account of those trees of the country whose flower, fruits, buds, leaves, roots, etc. are used as food or medicine. If according to the books of the Hindus, a man were to collect only one leaf from each tree, he would 18 get bars (or loads) (5 surkhs = 1 masha; 16 masha = 1 karg; 4 kargs = 1 pal; 100 pals = 1 tula; 20 tulas = 1 bar); i.e. according to the weight now in use, 96 mans…[55]

With the growth of urbanisation, the construction work, especially of durable buildings, gained momentum and wood constituted an important item in construction work as the windows, the doors and in some places entire houses were made of wood, for example in Kashmir.[56] About the use of wood in the houses in Kashmir Abul Fazl writes:

The houses are all of wood and are of four stories and some of more, but it is not custom to enclose them… On account of the abundance of wood and the constant earthquakes, houses of stone and brick are not built…[57]

Bernier writes that timber was readily available on account of its cheapness and the facility with which it was brought from the mountains by means of the so many small rivers. The wood was preferred for building of houses in many regions.[58] Pelsaert mentions that houses in Kashmir were made of Pine-Wood.[59] Jahangir too writes that the buildings of Kashmir were all of wood and were two, three and even four storied.[60] In other parts of the country too, wood was utilised for construction of the buildings for instance in Ahmedabad, the shops were usually made of wood.[61]

According to Abul Fazl generally eight kinds of wood was used for house building. Those were:[62]

S. No.	Wood	Size	Cost
1.	Sisau (Unrivalled for its beauty and durability).	A block 1 ilahi gaz long, and 8 Tassuj broad and high, if height be only 5 or 6 t.	15 d. 6 j 11 d. 10 ¾ j
2.	Nazhu (in Hindu Jidh)	A beam, 10 T. broad and high,	Per gaz 5 d. 13 ¾ j.
		A half size beam, from 7 to 9 T. broad and high.	Per gaz 5d. 3 ¾ j.
3.	Dasang or Kari	A beam 3 t. broad and 4 gaz long.	5 d. 17 ½ j.
4.	Ber	1 T. broad and high, 4 gaz long	5d. 17 ¾ j. so also tut or mulberry.
5.	Mughilan or Babul	Of the same cubic content as No. 4.	5 d. 2 j.
6.	Sirs	Size as before	10 d. 4 j.
7.	Dayar	Same size 1st quality	8d. 22 ¼ j
		2nd quality	8d. 6 ½ j.
8.	Bakayin	Same size	5d. 2 j.

Babur writes that the *Mahuwa* wood was widely used for house construction. According to him "*Most of wood in the houses of Hindustan is from it.*"[63] The wood of *Sirs* tree was also popularly utilised for construction purposes.[64] *Bas* or bamboos were used for covering ceiling.[65] The thatched roofs were supported by layer of strong canes.[66] Bernier writes in 1663 that more than six thousand roofs were destroyed in Delhi because of the fires for three times.⁶⁷ This indicates the use of the huge quantities of wood in the houses. In Gujarat pillars and ceilings of houses were made out of *saj-wood*⁶⁸

Wood was also largely utilised for the construction of bridges, Innumerable wooden bridges were scattered all over the country. During his journey to Kabul with along Akbar, Father Monserrate saw a wooden bridge being built over the Sutlej (in Machivara, Ludhiana district)⁶⁹ and noticed several others across the Beas (Dungarri), the Ravi (Kalanaur), and near Sodhra.⁷⁰ In the valley of Kashmir there were many strong wooden bridges (*kadals*), but they were susceptible to damage and destruction by fire and flood. Ample precautions were required to ensure their safety.⁷¹ Bernier saw two beautiful wooden bridges in Srinagar.⁷² Jahangir comments on the construction of two wooden bridges in Kashmir,

*They were both 5 cubits wide. In this land the method of building bridges is as follows: Trees with branches are laid across the surface of the water, and the two ends are fastened tightly to the rocks. Then thick wooden planks are laid on the top and fastened with pegs and strong ropes. With a little repair they last for years.*⁷³

Jahangir himself ordered the construction of four extremely sound bridges (*kadals*) of stone and wood in the city of Srinagar.⁷⁴ About the wooden bridges of Kashmir Inayat Khan writes:

Spanning the river there are 30 substantial wooden bridges resting on huge pillars, and four of these are situated in the city. Each of the bridges is the wonder of the age, and their study supports are not the least affected by the crossing of enormous elephants... ⁷⁵

In other parts of the country, too there were numerous wooden bridges. Streynsham Master reports in his diaries the destruction of some wooden bridges in a storm at Masulipatam in 1679.⁷⁶ At Madras, the French replaced the wooden bridges linking the town to the fort with a cause way.⁷⁷ These wooden and boat bridges were widely constructed and were so strong that the army with elephants, camels, horses, baggage and artillery crossed safely.[78]

The *sygwan* or teak, afforded the best timber for building in whatever branch. Those who built houses of the 1st class teak, rarely failed to built all the terraces upon teak joists; because they possessed superior strength, and that they were less likely to attacked by white ants. The greater part of teak was used in Bengal and at Madras.[79] The wood of a tree called *Talipot* or *Talpat,* common in the islands of Ceylon and coasts of Malabar and Coromandel was used for making the rafters for the buildings.[80] Inayat Khan writes that wood was widely used for construction purposes in Kashmir because of its large availability.

All the dwellings and buildings of the enchanting valley, with exception of those belonging to the Emperor, the illustrious princes and nobles, and also some of the opulent inhabitants of the country, are formed entirely of timber and planks.[81]

Though the wooden palace, except in Kashmir were no more thought of, the use of timber occurred on several occasions in the structures of the Medieval period in India. Roofing, doors and windows were commonly made of timber and wood.[82] Abul Fazl has appended a list of the varieties of timber along with their prices in the *Ain-i-Akbari,* which were available in the Agra market.[83] Similarly much later in c. 1825 Maulvi Khairuddin Lahori in his 1st volume of *Ibaratnama* recounted the most suitable timber for various purposes in the process of house building in the *Subah* of Lahore. The construction work specially of the durable buildings had grown enormously as the growth in urbanisation gathered momentum, therefore the consumption of timber in the buildings of the empire could not have been inconsiderable.[84]

The timber yield from the forests was also used in the building of the means of transport such as ships, boats, carts, palanquins, etc. for conveyance over land and sea routes. Aside from the porters and the beasts of burden, these being the only means of transport of goods then, plentiful supply of timber hence of transport vehicles was a great asset in improving and raising the frequency of transportation of merchandise from place to place.[85] While carts and carriages seem to have been built locally and every where, but in the 16th century timber was the determining factor in the ship and boat building and the ship and boat building industries were located in the places which on one hand, were within easy reach of the forests tracts and on the other hand an easily accessible to watery courses in order to ensure the conveyance of turned out vessels to the place of requirement[86] Kashmir, Lahore, the Western coast, Allahabad and Bengal were the major regions, which supplied raw material for this industry.[87] The town of Wazirabad due to its location upon the main highway between Lahore and Attock and on the account of its proximity to both the river and the land routes, witnessed the growth of a boat-building industry in the 17th century.[88] Boat building was also one of the principal industries of Multan. The great amount of the trade carried by river from Multan would have required a sufficient number of boats. A large number of boats were built further north (at Lahore and Wazirabad) where procurement of wood was easier, and from there these were then floated down.[89] Teak timber was understood as good for the manufacturing of ships.[90] Teak was mainly supplied from Western *ghats* as it was within the reach of the sea coast where large vessels were built.[91] Abul Fazl has regarded Bengal, Kashmir, and Thatah as the principal centres of boat building.[92] Abul Fazl mentions that 40,000 boats were used for transport at Thatah.[93] Jahangir writes that in Kashmir there were 5,700 boats and 7,400 boats men.[94] Large ships were built at Allahabad and Lahore and were sent to coast. In Kashmir a model of a ship was made which was much admired.[95] Bengal and Orissa supplied Bamboos for fitting out ships.[96] In Bengal bamboos were largely employed to fit out smaller vessels for movement within the *subah*.[97] Numerous varieties of boats were produced such as *ulak, bajra-a pleasure* boat, *purgoo, bhar-a floaty light* boat, *patella*, flat bottomed boat of exceeding strength and burden, writes Bowrey.[98] Even among the flat-bottomed boats there were separate categories which were constructed differently; each to suit the peculiarity of the rivers at different places. The wood used for their construction was obtained from the northern hills since the type and quantity of timber required was probably not available in the Plains.[99] Sujan Rai has mentioned that wood was floated down the Chenab from Chamba to Wazirabad which was a town engaged in the boat building.[100] It is possible that timber was also floated down the other rivers and that a fair amount of it was used for boat building.[101] The largest category of the boats named *doondah*, were found in the lower reaches of Indus in the region of Sindh, Further north, upon the upper Indus and Chenab, the *zoruk* was more commonly used boat. The *beri* might also be mentioned here. According to Baden-Powell, it was,

the largest traffic boat on the Ravi, Chenab and Jehlum and also on the Indus," and *"carries from 500 maunds or less up to 1,000 maunds.*[102]

There were some other boats such as *ghurrab* and *sabnak*. The Imperial *nawara* in Bengal was composed of 1000 *ghurrabs* which were commonly employed boats of the Great Mughals. *Sabnak* was a small delicate boat used for accompanying large vessels.[103] Kashmir was also famous for the manufacture of the different types of boats such as *Shikaras, Parindas* and *Takht-i-rawans*.[103a] *Sisso* wood was employed for the frame, ribs, knees, etc. of ships, especially those of great burden; for such it was found to be fully as tough and as durable as the best oak.[104] As a matter of fact India is said to have had the monopoly of the ship building trade as far as the lends washed by the Indian ocean were concerned; Indian built ships reached the ports of Europe. So ship building was a thriving industry, this was because the forests of India provided the cheapest material for ship building. The Arabs used ships built in India for their trade in Indian Ocean. Ship building continued to be thriving industry through out the Medieval period. In the 14th century Friar Oderic crossed the Indian Ocean in an Indian Ship manned by Indian Sea-men. In 1670 East India Company's factor at Balasore wrote to the court of Directors at London:

Many English Merchants and others have their ships and vessels yearly built. Here is the best and well grown timber in sufficient plenty, ... very expert master builders there are several here, they build very well and launch with as much discretion as I have seen in any part of the world. They have an excellent way of working shrouds, and any other rigging for the ships

Shipping remained an important industry up to the end of 18th century. The industry was destroyed because the ship builders of London raised a storm of protest to forbade the use of Indian ships in order to save its own ship building industry from competition.[105] Apart from ships and boats, timber was widely utilised in the manufacture of the other means of transport such as carts, *raths* (chariots), palanquins, etc. Palanquins or *Palki* was the most common means of land transport used especially by the upper class.[106] In Gujarat *Sheesham* wood was used in construction of *raths* and other means of transport.[107]

The botanical products were also utilised for the medicinal purposes. Almost all the sources of Mughal period throw light on the medicinal properties of the botanical products. The old kernels of mangoes were used as medicine.[108] Coconut too had many medicinal properties. A kind of coconut was used for the preparation of an antidote against poison *(Taryag-i-zahar)*.[109] The nut of the coconut is full of a sweet water, a drink that was very refreshing. This water of coconut was used in inflammation of the liver, the kidneys and the bladder and helped to increase urination. It was also good for excessive heat of liver, pains in bowels, or discharging of mucus or of blood. It also helped to refresh in the season of great heat.[110] Oil was also extracted from the nuts of coconut and was usually applied by the women to their hair, which provided greatest delight. The Oil of coconut was used for many medicinal purposes, such as in burns and ulcers. An ointment was also made with the coconut oil. It served for a purge to the lean and irascible, expelled bile, and reduced adipose tissue.[111] Fruit of a class of Palm tree – *Palmeira brava* (the wild palm) possessed many medicinal properties. Inside each fruit there were ordinarily three lumps of soft pulp which were very refreshing. Eating them was useful in clearing the sight of those unable to see at night.[112] The distilled juice of *ananas* (pine-apple) was useful for dissolving stones in kidneys and bladder.[113] The *Mirobalanos quebulos* (chebulic myrobalan) were used as medicines.[114] The *coquinhos* (little coconuts) were used in many complaints such as diarrhea and mouth sores.[115] The medicinal books of the period describes the advantages of the *amla* fruit which one could observe by eating of this wholesome fruit.[116] Rose water also served many medicinal properties. In hot weather people used to pound their bodies with rose water, sandal wood and also many scented and cooling oils to keep them cool.[117] In Gujarat, writes Barbosa, people anointed themselves with white sandal wood paste mixed with saffron and other scents to keep them cool.[118] Rose water was also utilised in preparation of cough medicine. A *tola* of wine, according to Jahangir, when mixed with rose water and water made a good cough medicine.[119] The bark of the roots of egg plum or *bair* tree (ziziphus jujube) was utilised in the preparation of medicine for frightful severe disease – the tapeworm.[120] The leaves of guava tree were utilised for medicinal purposes.[121] The fibers of lime boiled in water were used to cure a person who was poisoned.[122] The juice of kaju (cashew-nut) was used to clear all obstructions in the breast.[123] Lemonade was used for the cure of prickly heat during the summers.[124] Distilled water of Egyptian willow *(bedmus, bed-i-Mushk)* was very comforting against all the fevers caused by heat.[125] The Plantains were used by the physicians in India, as dressings for blisters, or as a covering for the shaven head in cases of brain fever.[126] The sherbet prepared from pumple nose (citrus decumanus) was a most grateful drink to the sick. The bark or hard external skin of Noona fruit (annona Reticulata) was a powerful astringent or tonic and was of great use in native medicines.[127] The Jamrool fruit (Eugenia alla or aquea) was regarded as a cooling medicine in the hottest months of the year and its bark was thought as a sovereign remedy for aphthae in children.[128] Tamarind or *ambli* also have the medicinal properties. Edward Terry writes:

... as the *tamaerine* tree, which bears cods some what like our beans, in which when the fruit is ripe there is a very well tasted pulp, though it be sour, most wholesome to open the body, and to cool and cleanse the blood.[128a]

From the tree products several house-hold articles were made. For instance, several domestic wants of the people were supplied by the coconut tree. A house could be entirely built of it. The walls and doors were made of cadjans (the leaves plaited), the roof was covered with the same, the beams, rafters etc were made of trunk. The builder needed no nails, as he used coir ropes made from the outside husk of coconut.[129] The shells (*pawast, cherata* or *chiratta*) of the cocos were used for making dishes, *(porzulanos)*, vessels, spoons, cups *(pyalas)*, handles, rings, rosaries and buckets to draw water.[130] Pans, drinking vessels, *hookah*, bowls, lamps, etc. were also made from the shells and brooms were made from the ribs of the leaflets. The refuse of the kernel, after the oil is pressed, served food for cows and pigs.[131] The shells were also used to make musical instruments such as *kasa-i-ghichaks* (a kind of violin).[132] Also the black spoons *(qara qashuq)* were made from the shells and the large shells were used to make guitar bodies.[133] The ropes for ships and boats were made from the husks of coconut and also cord for sewing boat-seams were made from the husks.[134] From the rind of coconut, pipes and cables were manufactured.[135] The importance of the coconut tree can be realised from the fact that if a man had a few coconut trees in his garden, he never starved.[136]

From the branches of the palm tree, mats were made and those served as carpets. People thatched their houses with its leaves and also used these leaves as decorations in the festivals.[137] Also from the leaves of palm tree were made the fans, mats, sun-shades, small baskets and other curiosities.[138] The fans made from the *khajur* leaves were also used in the royal processions.[139] Fitch has described the Palm tree as the most profitable tree in the world. According to him the leaves were utilised to make thatch for houses and sails for ships and branches were used to make brooms, and the wood was utilised for the making of ships.[140] Palm leaves were also used as plates (dishes) and small vessels for serving food were also made from them.[141] The plantain leaves were also used as plates and dishes for serving food.[142] After twisting and curling the leaves of tar, a sort of ornaments for the ears of women were made, those were also sold in the markets.[143] From the bark of San (hemp) plant, strong ropes were made and from another variety of this plant called Pat-san, very soft ropes were made.[144] In Bengal people made mats from the branches of *tesi* tree.[145] The leaves of Talipot or *talpat* tree were used for making umbrellas. The leaves were folded up in plaits, like a fan, and were cut into triangular pieces, which were used as umbrellas for protection against sun or rain. These leaves were so large that when these were hung from the top, one of them being sufficient to cover 15 to 20 men. The tents of the *Kandian* Kings and others, in the times of war, were made of these leaves, and hence were called *Tal-ge* or *talpat* houses. These leaves were also used to cover carts, palankeens, or any thing that was necessary to keep from the sun or rain while traveling.[146] The wood of old guava trees is exceeding close grained and tough. It was much used amongst the natives of India for Gun-stocks, it took a great polish, and was rarely known to split with heat or fracture from blows. The wood of mango trees was also most extensively used, and in fact the planks, supply for a larger part of India. When carefully preserved by paint it lasted for many years. From the leaves of pine-apple plant, an exceedingly beautiful flax, of great fineness and strength was prepared by simple maceration and beating.[147]

Wood work of excellent quality was done through out the country. A number of House hold articles were made. For instance, doors, peg, seats, toys, bed steads, and various other implements and vessels were made of wood.[148] The Wood of Sisso tree was utilised for the construction of many house-hold articles such as furniture especially chairs, tables, tepoys (or tripods), bureaus, book-cases, escritoires, etc.[149] Kashmir was famous for its excellent wood work.[150] Bernier has regarded the wood workers of Kashmir as very active and Industrious.

The workmanship and the beauty of their palekys, bed-steads, trunks, ink-stands, boxes, spoons, and various other things are quite remarkable, and articles of their manufacture are in use in every part of India. They perfectly understand the art of varnishing and are eminently skillful in close imitating the beautiful veins of certain wood, by inlaying with gold thread so delicately wrought that I never saw any thing more elegant or perfect.[151]

The carpenters of Kashmir were famous for their adroitness and excellence. Excellent specimen of their lattice work (*tabdan tarashi*) are extant in the mosques of Shah-i-Hamadan, Bahauddin Sahib and Madin Sahib. The art caused much wonder to Mirza Haider Dughlat and Bernier. The carpenters of Kashmir also rendered themselves famous as builders of ornamental ceilings (*Khatam bandi*) and as wood-carvers, cabinet makers and builders of wooden bridges.[151a] De Laet has also made reference to the manufacture of the wooden boxes in Kashmir.[152] Praising the wood workers of the country Edward Terry writes:

Their skill is likewise exquisite in making of cabinets, boxes, trunks and stand dishes curiously wrought…They paint staves or bedsteads, chests of boxes, fruit dishes…extremely neat which when they be not inlaid as before, they cover the wood, first being handsomely turned, with a thick gum, then put their paint on…and after make it much more beautiful with a very clear varnish put upon it.[152a]

In Bengal and Orissa, several varieties of bamboo was found which were used for making all kinds of furniture.[153] De Laet records that

tables, tablecloth and napkins which the people used were made of plantain leaves[154]

Leaves of the trees were often utilised for the writing purposes. Babur records that *Tar* (palmyra Palm) leaves, which were about a yard in length were used by the people to write Hindi characters after the fashion of account rolls (*daftar yu sunluq*).[155] In Kashmir many Brahmans used to write on the leaves called *Bhoj pattra* and on the bark of *Tuz* tree. All manuscripts were written on them.[156] *Tar* leaves were used by the people of Orissa for writing purposes. Iron pen was used for writing on these leaves.[157] Early travellers notice that in Southern India all writing was done on the Palm leaves, and as late as 1625, Della Valle obtained a specimen manuscript, it was written for him on this material.[158] The leaves of *Talpat* tree were used in strips to teach the children to write upon, and as every letter was cut into it by a sharp pointed style, the writing was indelible and continued legible as long as the leaf itself lasted.[159] Leaves of the trees were also used for maintaining records for instance in Patna (*Sarkar* Mungir) there was a temple Baijnath. In this temple there was a *pipal* tree. The leaves of this tree were utilised for keeping records. The text mentioned the amount that was given to some person by the temple. Even if the house of the given person was 500 *karoh* away from the temple, his name and the names of his wife and children were written on the leaf correctly. The servants of the temple used to keep the leaf and also gave a copy of that to the person concerned. It was called *Baijnath-i-Hindi*.[160] Barks of certain of certain were also utilised for making paper for instance, 'white paper' of Bengal was reported to have been made from the bark of a tree and was 'smooth and glossy like a deer's skin'.[161]

From the frequent references in the account of the European travellers one finds that spirits and fermented liquors were easily procurable through out the country.[162] There were some trees from which intoxicants were extracted. From the flowers (*gilaunda*) or *mahuwa* intoxicating spirit *(araq)* was extracted.[163] Peter Mundy, Mandelslo and Ovington comment that the most common and perhaps the cheapest drink was *Tari* or Juice of Coconut Palm or date trees, which were pleasant in taste and flavour and was popular through out India.[164] From the *tarkul* or palm tree, *tari* was received three times a day.[165] Delaet mentions about the extraction of *Tari* or *Suren* (*Sura*, liquor) from trees at Broach.[66] *Tari* liquor was more exhilarating than date liquor. People used to hang a pot on the tree and took its juice and drank it. [167] Tari juice was also called *Sura* and *Nira*.[168] The date palm *(Khurma)* yielded another intoxicant.[169] If it was drunk fresh it was sweet and pleasant, but if it was kept for a day or more, it was quite exhilarating *(Kaifiyat)*.[170] About the juice of date palm Babur writes:

…excursion to the villages on the bank of Chambal river, we met in with people collecting this date liquor in the valley bottom. A good deal was drunk, no hilarity was felt, much must be drunk, seemingly, to produce little cheer.[171]

Tavernier mentions that in Golkunda women ran the *Tari* shops.

The King of Golkunda derived huge amount from the tax imposed on tari sale.[172]

Fitch, William Finch and Edward Terry also mention that a kind of intoxicating liquor was drawn from the date and coco tree which was called *"tarrie"(tari, toddy)* or *"sure"* (Sanskrit, *sura*). Terry writes that if it was taken fresh it was very tasty, but if it was taken after few hours, it became heady, ill relished and unwholesome. Terry also describes it as piercing medicinal drink,

If taken early and moderately, as some have found by happie experience, there by eased from their torture inflicted by that shame of physicians and tyrant of all maladies, the stone.

Terry also mentions about a kind of wine which was made from sugar and spice rinds of a tree called *jagra*. *Jagra* was a coarse sugar made from the sap of various palms.[173] About the liquor obtained from the coco tree, Careri *says*

it is so nourishing that Indians live upon it without any sustenance.[174]

According to Manucci:

When drunk fresh it is sweet and suave, but if kept for 12 hours, it tastes like beer and goes to the head. It is used by the Indians in place of wine. From this liquid are manufactured aqua ardente, vinegar and sugar.[175]

Coco Juice was the principal ingredient for the preparation of a liquor which "drinks as deliciously as wine". Indians particularly the Goanese, liked it very much and drank it like water. It was very strong especially after the third distillation.[176] Sugarcane was also used for the preparation of intoxicating liquor.[177] Nira was another kind of wine drawn from arequier tree and was sweet like milk.[178] Nicolo de conti speaks of the great prohibitionist country of India

for throughout all India there are no wines, neither is there any wine,

and flatly contradicts himself by saying

...but they make a drink similar to wine of pounded rice mixed with water, the juice of certain trees, of red color, being added to it.[179]

However, Nicolo de conti's statement about being no wine in the country can not be regarded as true. Probably he gave a specific meaning to wine, that is to say, wine made of grapes. The fact is that the Indians were addicted to wine. Poor class people used to drink toddy, the juice of a kind of palm tree and rich men used to enjoy the delicious blend of Shiraj, a place in Persia.[180]

Certain botanical products were also used as dyeing and mordanting agents in the Textile industry. The cultivation of crops yielding dye stuffs was greatly helpful in the progress of the textile industry. Besides from the annual cultivation of indigo, al and so on, there were several other trees such as lac, tun, catechu, bel and others, occurring in the various parts of the country from which colouring materials for various shades were obtained.[181] Many flower plants yielded dyeing contents such as *henna* and saffron. These dyeing agents were used for dyeing cloth, paper, utensils, hand, feet and hair.[182] The juice of *ratanmala* flower when boiled and mixed with vitriol and *musafar*, furnished a fast dye for stuffs. And when the butter and sesame oil were boiled together with the root of *ratanmala* plant, purple dye was obtained.[183] The bark of chanpala plant was boiled with water to get red dye.[184] The lac also yielded red dye and served as sealing wax and varnish. The lac produced in Bengal was the best, cheapest, and the most abundant.[185] Tavernier writes that the lac was dearer in Bengal. People extracted beautiful scarlet colour from it which they used to dye and paint their cotton cloths.[186] Tavernier mentions that Assam also produced abundance of shellac of two kinds. The shellac that was formed on the trees was of the red colour and with it people used to dye their calicos and other stuffs.[187] The flowers and fruits of pomegranates were also of great value to dyers, who extracted the dyeing colour from them.[188] The bark of Noona fruit was also utilised in the dyeing processes.[189] Besides this Safflower was also used as dyeing agent. It was abundantly produced in the provinces of Bengal and Bihar and it was easily available in bazaars as may be adjudged from its frequent use by the dyers. Besides other mordanting agents such as lemon and mangos were grown widely in these regions.[190] These dyeing agents enabled the Indian fabrics to acquire better finish, which enhanced their market value. Further the multiplicity of these colouring agents offered a wide range of choice to the Indian dyers and traders so that the residue, usually the expansive indigo, could be exported abroad.[191]

Mulberry trees facilitated the emergence and expansion of sericulture in India. Bengal[192] and Kashmir [193] were the main territories where mulberry trees were grown on large scale. In Kashmir mulberry trees were cultivated on the large scale. Kashmiris imported silk worm eggs from the neighbouring provinces of Gilgit and Little Tibet.[194] and nurtured them on local mulberry trees.[195] This import had, on the one hand, improved the quality of Kashmiri silk, on the other hand stimulated the silk industry to such an extent that Mirza Haider Dughlat regarded its enormous volume as one of the wonders of Kashmir.[196] Evidently it was on account of the superior quality and sizeable quantity of this silk, Emperor Akbar reserved it as an imperial monopoly.[197] Jahangir also mentions the rearing of the silkworms on the mulberry trees in Kashmir to facilitate the manufacturing of silk.:

In fact, the mulberries of Kashmir are not fit to eat, with the exception of some of the trees grown in the gardens, but the leaves are used to feed the silk worm. They bring the silk worm's eggs from Gilgit and Tibet.[198]

The Bengal mulberry trees were of small stature just about two or three feet in height.[199] The silk worms reared on these trees were of four varieties: *nistari, desi, harapalu,* and *china palu*.[200] *Ghoraghat*[201] and *Maldah* were reputed for both rearing of silk worms and production of silken stuffs.[202] Tavernier has given an estimate of the volume of silk production in Bengal. He writes that Qasim bazaar in Bengal alone could furnish about 22,000 bales of silk annually, which at his equitation of a bale with 100 livers might mean 2.4 million *lb.avdp*.[203] He writes that carpets of silk with gold and silver were manufactured at Surat.[204] He has also mentioned about *patolas*, a kind of silk, very soft, decorated all over with flowers, manufactured at Ahmedabad.[205] By about 17th century the silken stuffs fabricated in Bengal were of considerably fine texture and the embroidery too was exquisite.[206] But the Bengal Silk was not comparable to that of Persia or Syria.[207]

Besides the true or mulberry silk, there were other semi-domesticated or semi-wild varieties which must have accounted for a large portion of Indian silk production.[208] In Bihar and Orissa, *'tassar'* producing trees yielded substance which led to the process of texture resembling the silk.[209] It was also produced in some parts of Deccan.[210] Fitch has also mentioned about the production of Herba (Tassar) silk in Orissa.[211] Bengal also produced this type of silk. There are notices of the herb silk in the accounts of foreign travellers. Pyrard speaks of the silk herb; Linschoten speaks of a kind of cloth spun from the herb and Caesar Frederic calls it the cloth of herbs, "*a kind of silk which growth among the woods.*"[212] *Eri* silk was produced in Bengal, Bihar, and Orissa.[213] And *'Muga'* and *'champa'* in Assam.[214] About the silk production in Assam, Tavernier writes,

There is a kind of silk which is produced on trees, and is made by an animal like our silk worm, but it is rounder and remains for a whole year on the trees. The stuffs which are made of silk are very brilliant, but soon fray and do not last long.[215]

Another important item which were extracted from the flower and the plants was the scented oils. Odoriferous plants of both foreign and Indian origin were used for extracting scented oils. Abul Fazl has listed a dozen varieties of perfumed oils, briefly describing the ingredients used and the method employed.[216] He has also given a long list of flowers which were used for making perfumed oils such as *sewti, bholsari, chambeli, mongra, champa, nargis,* etc.[217] The more sophisticated perfumes too were extracted from flowers, having stronger and more refined scents such as that of roses, *gul-i-henna*, jasmine, saffron and so on.[218] Babur writes that excellent perfume was extracted from the *Kewra* flower.[219] Precious scents of the diverse kinds were in use in India from the ancient times. Kautilaya's *Arthashastra* gives a long list of fragrant substances for toilet preparations.[220] Malik Muhammad Jaisi makes a particular mention of two strong scents or *ottor* named *maidu* and *chuvi*, but their specific variety is not clear.[221] The *Ain-i-Akbari's* account of scents is no less detailed and their prices ranged from half *rupee* per *tola* for *zabad* to 55 *rupees* per *tola* for *sandal wood*.[222] *Araq-i-Sewti, araq-i-Chameli, mosseri* and *amber-i-ashab* were considered best among the different varieties of perfume.[223] Cultivation of sweet scented flowers like roses of various kinds such as *bela*, jasmine and many others were under taken by extracting rose-scent, rose oil, rose water, *itr (uttor)* from them. Jaunpur, Ghazipur, Jalesar and Qannauj were the chief centres of the manufacture of these items. Even today also Jaunpur holds the field in spite of market being flooded with foreign products and is noted for perfume essence and fragrant oil, chiefly *bela* oil and *itr* of Qannauj were renowned through out the country.[224] From roses diverse kinds of perfumes were extracted, the most notable being *Jahangiri itr*, which was invented by Nurjahan's mother, Asmat Begam.[225] But Manucci has attributed the discovery of this perfume to Nurjahan.²²⁶ Large tracts of the land were brought under the cultivation of these odoriferous plants around Champaner, Ahmedabad,²²⁷ Surat ²²⁸,and Sironj²²⁹ and perfume making industry developed in these towns as also at Nausari.²³⁰ Popular fragrances were made from amber, camphor, *chuwa* (aloe), ud or wood or aloes, sandal wood and *sugandh gugala* (bdellium).²³¹ In Champaner the perfume was extracted from the aloe wood called *mandali*, the best variety of ud. Its powdered form was used for rubbing into skin and clothes.²³² Sweet scented oils of diverse kinds were extracted. In Bengal, people anointed their bodies with sandal²³³ and other oils extracted from various flowers.²³⁴ In Gujarat, according to Barbosa, people anointed themselves with white sandal wood paste mixed with saffron and other scents. ²³⁵ In hot weather rich rubbed their bodies with pounded sandal wood and rose water, and some other scented and cooling oils.²³⁶ *Santak* – a product of civet, *chuwah*, *chamebli's* essence and rose water was also used for the same purpose.²³⁷ A large amount was spent on the manufacture of scents from the flowers by the royal family. According to Manucci:

*the expanses of the mahal are extraordinary, for they never amount to less than a carol(Karor) of rupees – that is 10 millions of rupees…the above expenditure will not appear incredible when we consider that all persons in India being extremely choice about, and fond of, scents and flowers, they disburse a great deal for essences of many kinds, for rose water, and for scented oils distilled from different flowers.*²³⁸

Botanical products were utilised in several other ways also. For instance Gum-lac was used for the manufacture of the women's bangles and toys.²³⁹ Tavernier writes that gum-lac in Bengal was used for making toys and sealing-wax…

*That which remains after the colour is extracted is used only to embellish toys made in the lathe, of which the people are very fond, and to make sealing wax; and be it for the one or the other purpose they mix whatever colour they desire with it.*²⁴⁰

Dutch exported large quantities of the Gum-lac to Persia, where it was used to produce the colours, which Persians employed in their dyes.[241] In Surat women prepared the sticks like Spanish-wax from lac and gave them colours whatever they wish. These were exported by British and Dutch companies in the large quantities.[242] Gum-lac was also used for varnishing furniture, doors, windows, etc.[243] Saffron was another product which was used for various purposes. This fragrant and costly article was used for imparting fragrance and agreeable yellow colour to some rich dishes. It was also used for dyeing cloth. The Rajputs had the custom of putting a *kesria* dress(dress dyed with saffron) on certain ceremonial occasions. It was an article of luxury mostly in demand by the rich class of people.[244] Saffron was produced in Kashmir.[245] Kishtwar variety was considered as the best. The demand for saffron was so great that it was not available for export.[246]

Botanical products were put to numerous other uses also. For instance, the ashes of *mughilan* (Babul) wood was used for refining the adulterated silver.[247] In Kashmir the bark of *Bar* tree was used for collecting gold from river Bihat (jehlum). Sujan Rai writes:

In the fords of the river the people spread the bark of the Bar tree around which they place stones, so that the current may not sweep it away, after 2 or 3 days it is taken out, left in the sun, and when dry shaken, yielding gold up to three tolas.[248]

San hemp, a plant was used to make ropes for well buckets; peasants also mixed it with quick lime.[249] *Khas*, a sweet smelling root of a kind of grass which grows along the banks of the rivers, was used by the people to make screens during summers and these screens were placed before the doors and were sprinkled with water. These screens rendered the air cool and perfumed[250] *Luk* was the flower bunch of the reed which was used for matting. People burned it and used it as candle.[251] Bamboo was also used for spears. A peculiar kind of the bamboo was also used for making thrones.[252] *Patal*, made of reed was used for *qalams* (pens). It was also used for covering the ceiling.[253]

Sikri was used to adorn ceiling and the walls of the houses.[254] *Munj*, the bark of the *Qalam* reeds was used for making ropes to fasten the thatching.[255] The best produce of the Egg-plum tree (*bair*) was the strong and durable silk called Tusser.[256] From the rubber and cinchona plants, quinine was extracted, which was of great economic significance.[257] Various kinds of dry fruits such almonds, raisins, *caju* were used in the preparation of various dishes.[258] Moreover due to the increased production of the fruits, the fruits were also conserved. In Bengal people preserved citrons in large quantities[259] From Bernier's account we find the similarity between the method employed in the preservation of citrons in Bengal and that of Europe.[260] In Thaneshwar, the conserves – *murrabba, chutneys* and *achars* (pickles) of vegetables and fruits were a flourishing industry. Mangoes, myrobalans (*hur*) were the chief articles of the conserves.[261] Manucci also mentions about the preparation of the mango pickles.[262] Abul Fazl has given along list of the pickles that were prepared from mangoes, sour limes, lemons, apples, raisins, *karil berries*, etc[263] A kind of sealing wax was prepared from the gums (sap) of the *Kunar*, the *Bar* and the *Pipal* trees. Like wax it also got warm when exposed to fire, but after cooling it becomes hard. It was mostly used to seal the royal *farmans*. The *farmans* were put into golden bags and were sealed with the help of this wax.[264]

Lac (varnish work) was used for *chighs* (*chiq*) (sliced bamboo sticks, placed horizontally, and joined by strings, with narrow inter sticks between the sticks. They were painted and used as screens). For red varnish, 4s. of lac was mixed with 1s. of vermilion; for yellow, 4s. of lac, 1s. of *zarnikh* (*auripigment*), for green 1/4s. of indigo was mixed with lac, and *zarnikh* was added and for black, 4s. of lac and 8s. of indigo was mixed.[265] Gum lac was also used by the gold smiths for filling in the ornaments.[266] From sugarcane, sugar and *gur* were manufactured. Different varieties of sugar was made.[267] The most refined and esteemed form of the sugar was the crystallised white *qand* (sugar). The manufacture of sugar was carried on a fairly large scale in Hindustan. In Bengal, sufficient was produced to leave a good surplus for export after local and internal consumption. Bengal manufactured granulated sugar and prepared various candied and preserved fruits.[268] In fact Bengal was the principal seat of the industry, the product was "powder sugar", which probably was the fine-grained type. This type of the sugar was also procurable in Ahmedabad. But most costly form called candy was produced in Lahore. The difference in the value between two types was considerable. Abul fazl gives the price as 128 dams for a *maund* of powdered sugar and 220 dams for a *maund* of sugar candy.[269] The sugar was universally used all over the country as shown by the numerous descriptions of sweets and sweet dishes and can be gathered from the mention of the sale of sugar and sugared drinks, in contemporary literature.[270] Tavernier declares:

even in the smallest villages… sugar and other sweet-meats dry and liquid, can be procured in abundance.[271]

Punjab also produced a fairly large amount of sugar. Pelsaert informs us that *"very much sugar is produced"* in Punjab. He refers to the production of sugar both in Lahore and Multan.[272] Thevenot records it as one of the exports of the *Suba* of Multan. He further adds that the Subha of Lahore yielded the *"best sugar of all Hindustan"*.[273] Manucci, too was highly impressed by the amount of the sugar produced in Lahore region.[274]

A kind of Saffron was produced in Surat also, but it was only used as colouring agent.[275] In Assam and Bengal, from the ashes of the leaves of the fig tree, the lye was made to boil silk. With this method the silk was made as white as snow. Tavernier writes:

…if the people of Assam had more figs then they have, they would make all their silks white, because white silk is much more valuable than the other, but they have not sufficient to bleach half the silks which are produced in the country.[276]

The use of trees for shading purposes was very much appreciated during the medieval period. For this purpose a number of fruits and shady trees were planted on both sides of the road either directly by the state or else by some wealthy people. Such shady trees were known as *khayaban* or avenues.[277] The whole road from Agra to Lahore was planted with shady trees on Akbar's order.[278] Jahangir also got the trees to be planted on the roads from Agra to Bengal.[279] Manucci has given the account of plantation of trees on the road from Multan to Allahabad.[280] William Finch also mentions the plantation of trees on the road side from Kabul to Agra.[281]

Botanical products not only fulfilled the needs of the people, but also were the source of income to various artisans and other group of people They provided occupation for a number of crafts namely weavers, carpenters, dyers, wood workers, builders, perfumers and so on. In some places enterprising businessmen engaged a number of craftsmen to manufacture articles under their own supervision. Of such organisations or factories, the best equipped and most efficiently organised were known as *karkhanas* or workshops.[282] Traditionally, there were 36 *buyutat* or Karkhanas; some of *karkahanas*, mentioned in *Ain*, are fruitery, the perfumery, the kitchen, workshops for the manufacture of shawls and textiles, etc.[283] Numerous artisans specialising in different crafts were employed in royal Karkhanas, as is evident from Bernier's account of imperial *karkhanas* at Delhi;

Large halls are seen in many places called karkhanas or workshops for the artisans. In one hall embroiders are busily employed, superintended by a master. In other... in a fourth varnishers in lacquer-work, in a fifth, joiners..., in a sixth, manufacturers of silk, brocade and fine muslins.[284]

The royal factories at Delhi sometimes employed as many as 4,000 weavers of silk alone besides the manufacture of other kinds of goods for royal supply[285] The Dutch factory employed 700 artisans for the manufacture of the silk at Qasim bazaar. A number of the artisans expert in making silk were employed by the English.[286] Abul Fazl states that Akbar had studied the whole production of silk stuffs and under his care foreign workmen had settled in India, and silk-spinning had been brought to perfection.[287]

The textile industry including silk was a major component of the artisan production and probably accounted for between 20 and 30% of the total artisan population of towns and villages. Perhaps 60,000 specialist weavers and dependants lived in Benaras region, 250,000 in Awadh, 200,00 in Rohilkhand and 30,000 in lower Doab. There were three levels aristocratic consumption and for export by the East India company. It was centred on a few illustrious centres; in Bengal, Dacca, Murshidabad. Benaras was a famous for silk saris and brocaded work[288] The rulers of Benaras brought in weavers from as far away as Gujarat in order to expand their production of fine silks and brocades.[289] The 'Nagpuri' Gosains of Benaras who worked at southern route were noted as traders in silk, while weavers in near Azamgarh produced fine quality turbans for export to the Maratha rulers among whom the cocked turban was becoming a distinguishing mark of high nobility.[290]

The royal fruitery encouraged the cultivation of new improved varieties of fruits; the royal gardens were constantly busy in raising the quality of the local fruits and importing new varieties, for this purpose expert horticulturists and gardeners were employed on large scale. Abul Fazl writes that Akbar invited horticulturists from Iran and Turan and got them settled in India.[291] In this department, *mansabdars*, *Ahadis*, and other soldiers were employed, the pay of a foot soldier varied from 140 to 100 *dams*.[292]

With the development of urbanisation, the construction work gained momentum. In this department too a large number of worker were employed such as carpenters, *pinjara saz*, *Arrakash*, bamboo-cutters, *patal band*, *Lakhira*, etc.[293] Abul Fazl has given a detailed account of the wages of these craftsmen. 1st class carpenter received 7*dams*., 2nd class 6 *dams*., 3rd class 4*dams*., 4th class 3 *dams*. and 5th class carpenter received 2 *dams*. For plain job work, a 1st class carpenter got 1*dam*. 17 *jitals*. for one *gaz*, 2nd class, 1*dam* 6 *jitals*. and 3rd class got 21 *jitals*.[294] A *Pinjara-saz* was the lattice worker and wicker worker. His wages were as follows: 1st when the pieces were joined (fastened with strings) and the interstices be dodecagonal 24 *dams*. for every square *gaz*; when the interstices formed 12 circles, 22 *dams*; when hexagonal, 18 *dams*; when *jafari* (or rhombus like, one diagonal being vertical, the other horizontal), 16 *dams*; when *shatranji* (or square fields as on a chess board), 12 *dams*. for every square *gaz*. Secondly when the work is *ghayrwasli* (the sticks being not fastened with strings, but skilfully and tightly interwoven), for 1st class work 48 *dams*. per square *gaz*; for 2nd class work 40 *dams*.[295] Then there were *Arra-kash* who used to saw beams. Their wages for job work were; per square *gaz* 2½ *dams*., if *sisau* wood; if *nazhu* wood, 2 *dams*. A labourer employed for a day, 2 *dams*., there were there men for every saw, one above, two below.[296] Bamboo-cutters used to get 2 *dams*. per diem. *Patal-band* used to get *1dam*. for 4 *gaz*. And *Lakhira* who varnished the reeds, etc. with lac, got 2 *dams*. per diem.[297] Ship and boat building was another thriving industry, which gave occupation to a large number of artisans. A large no. of boats were built in several parts of the empire. Fitch travelled from Agra to Bengal with a fleet of 180 boats. The boats in use were fairly large; at Lahore they were 60 tons and upwards; some of the barges on the Jamuna were of 100 tons, while those on Ganges ranged up to 400 or 500 tons.[298] Ship building was an important industry in Akbar's time. Abul Fazl provides definite instances of ship being built at Lahore. He also writes that in Kashmir boatmen and carpenters derived a thriving trade.[299] He writes about the construction of four ships by the artisans on the orders of Akbar in his 38th regnal year (1598), *"Artificers by the command of His Majesty commenced to build four ships."*[300] The very next year a ship was supposedly completed on the banks of the Ravi. The length of the keel (*cobi*) which formed the foundation of this wooden house was 35 *ilahi* yards. 2936 large planks (*shatirs*) of *sal* and pine were used in its building and 240 carpenters and blacksmiths and other artisans were employed.[301]

Apart from those who worked in the Imperial *karkhanas*, there were other artisans who worked on their own initiative. In Batala many artisans were employed in the wood work.[302] Artisans of Kashmir were also famous for their wood work. Bernier *has regarded them very 'active and industrious'*. He has praised their workmanship and the beauty of their products. He writes:

They perfectly understand the art of varnishing, and are eminently skillful in closely imitating the beautiful veins of certain wood, by inlaying with gold thread so delicately wrought that I never saw anything more elegant or perfect.[303]

Sialkot by the end of 17th century emerged as town of considerable artisanal manufacturers. Its artisans produced famous *Man singhi* paper, embroidery work with silk and gold was also carried out there.[304] Scents and scented water industry also developed in various parts. A whole class of scent merchants, for instance, exited in Bengal and were known as *Ganda Baniks*[305]

There were numerous other professions related with botanical products in which several artisans were engaged. Tavernier writes that many women at Surat gained their livelihood by preparing lac after the colour has been extracted. They gave it colour as they wished and made it into sticks like Spanish wax.[306] He also mentions that in Golkunda the women ran the tari shops. The king of Golkunda derived huge amount from the tax imposed on the tari sale.[307] Sugar and gur manufacture constituted another important village industry, which was the source of income to many people.[308] During the courses of their travelling, the imperial court spent the nights and days in well-planned camps. These large tents were made of timber and in fact were portable buildings which could be taken apart and reassembled; for each of these two camps were employed about fifty carpenters along with other workmen.[309] Sujan Rai writes in the reign of Aurangzeb that in Kashmir, boats were the main mode of transport and were mainly used in transporting heavy articles. Hence boat-men and carpenters derived a roaring trade there.[310] There was a special class for rearing of silk worms in Bengal, who looked after them and some members of this class may still be found in Maldah, Rajshahi and Murshidabad districts. For winding of silk yarn from the pod generally a large number of women were employed who used to perform this work with such extreme delicacy that ordinary vision failed and only the exceedingly sensitive fingers could be of assistance in making out the difference in the various grades of its thickness (or fineness).[311]

Thus, the artisans who excelled in the manufacture of the products, were not only employed in the Imperial *karkhanas*, but their skills were also improved by these *karkhanas*. The craftsmen employed in the *karkhanas* were at pains to produce articles which in turn brought them rewards and promotions. The handicrafts practiced in the Indian sub-continent earned the admiration of the connoisseurs every where in the world. Mughals improved the standards attained under the Sultans by continuing their tradition of Patronage, contacts with the world had widened; the wealth of the empire had multiplied; peaceful progress in the neighbouring lands, especially in Iran, and conquest of Kashmir had given impetus to industries and skilled immigrants were encouraged.[312]

These Botanical products played an important role in the economy of the country. Brisk trade of these products was carried on the inter-provincial level and there was well established foreign trade also. Among the articles exported by Mughals were sugar, silk, candid fruits, perfumes, sandal wood, gum lac sealing wax, etc. A large number of articles were also imported to meet the needs of the people such fresh and dry fruits from Central Asia,[313] perfumes of Persia, silk,[314] Damishqi rose water, etc.[315]

There was a brisk inter-provincial trade in the botanical products. Different parts of empire excelled in different products such as Kashmir was famous for Saffron,[316] fruits and wood work.[317] Manucci writes:

In the province of Kashmir much fine (woolen) clothes made... They make many beds, ink-boxes, trays, boxes, spoons, etc.., out of wood both in plain and carved works. Fruits in plentiful...[318]

Kashmir was famous for its silk also.[319] Bengal was famous for its sugar[320] and silk.[321] In Aurangabad were produced sugarcane, mango, coconut, *keora* (leaves yielded the rich *keora* perfume).[322] Boat building was a thriving industry in Surat, Bengal, Thatta and

Kashmir.³²³ Jaunpur, Ghazipur, Jalesar and Qannuj were the chief centres of the manufacturer of the rose-scent, rose oil, rose water, etc. Jaunpur which even now holds the field inspite of market being flooded with foreign products was noted for perfume essence and fragrant oils, chiefly *bela* oil, *itr* of Qannuj was renowned through out the country.³²⁴ Surat³²⁵ and Nasauri in Gujarat were also famous for their perfumes.³²⁶ Kabul was known for its fruits.³²⁷ Kashmir supplied silk, saffron and fruits to different provinces. Kashmir had monopoly of the saffron production in India. The demand for saffron was so great that it was not available for export.³²⁸ Kashmir sent to Agra, saffron and fruits. Pelsaert writes:

*Kashmir yields nothing for export to Agra except saffron, of which there are two kinds. That which grows near the city sells in Agra at 20 to 24 rupees the ser; the other kind which grows at Casstuway, 10 kos distant, is the best, and usually fetches 28 to 32 rupees the ser (of 30 piece weight)... walnuts, which are –plentiful are also exported to Agra.*³²⁹

Lahore was linked with Kashmir through the passes in the Pir Panjal mountains and the river system.³³⁰ These indeed were the only convenient routes between Kashmir and the Mughal empire – a circumstance which rendered Lahore vitally important for Kashmir trade. Thus it received from Kashmir, silks, boats, saffron, dried raisins, walnuts, fresh fruits, timber, etc. These articles were thence distributed to their respective destinations such as Agra, Kabul and Multan.³³¹ William Finch has described the route from Lahore to Kashmir: *"Lahore – Gujarat - Bhimber - Peckly (pakhli) – conowa - Cassimer (Kaashmir)"*. He writes that Kashmir was abounding with fruits, saffron, etc.³³² Trade in wood was carried through river Chenab. Merchants used to bring *sal* and famous teak wood by the river highway from the hill country of Chamba and else where to Wazirabad, and made profit. Having built rafts with the wood, they used to carry them for sale along this river to Bahakkar and Thatta.³³³ Ahmedabad used to get **coconuts** from Malabar and sandal wood from Europe.³³⁴ Orissa annually exported by sea, Lac to the

Coromandel coast.³³⁵ Bengal exported silk to Coromandel up to Ganga, it also exported rice and silk to Patna. There was brisk trade on and along the Ganga and Jamuna up to Agra. Agra imported sugar and raw silk from Bengal and Patna. Patna seems to have been an important mart for Bengal silk probably owing to its convenient position with relation to Agra. From Agra again, sugar and Bengal silk were carried to Gujarat.³³⁶ From Bengal sugar was also carried to Coromandel and around Cape Comorin, to Kerala.³³⁷

Apart from the Inter-provincial trade, a brisk foreign trade was also carried in these products. India exported many botanical products to the countries like Japan, Pegu, Persia, etc. and also imported many products from them too. New materials like silk, sandal wood, etc. found their place in the foreign market. In Akbar's reign, there was only one foreign group which was supreme in the markets of India, namely the Portuguese; but in Jahangir's time the Dutch, the Portuguese, and the English were competing for the Indian goods. These traders also opened the trade with Japan and China and so the enter pot trade was increased.³³⁸ In the course of 17th century, an important sea-borne trade developed in Bengal silk, notably through the agency of Dutch, who exported it to Japan and Holland. They were said to have taken 6,000 or 7,000 bales yearly from the Qasimbazar market; they would taken even more if the merchants from the other parts of the Mughal empire and Central Asia had allowed them to do so.³³⁹ Apart from raw silk, many silken stuffs were supplied to the foreign lands. *Patoles* (a silk cloth), very soft, decorated all over with flowers of different colours, were manufactured at Ahmedabad. Their price varied from 8 to 40 rupees the piece. This was one of the profitable investments of the Dutch, who did not permit any member of the company to engage in its private trade. They exported it to Philippines, Borneo, Java, Sumatra, and other neighbouring countries.³⁴⁰ Silk was also exported to Middle East.³⁴¹

Fruit conserves of Ahmedabad and Thaneswar figured amongst the exportable consignments particularly to Arabia and Iraq, in weight its *maund* was allowed to 40 & 2 and a half seers.³⁴² The gum-lac of Bengal was brought by Dutch, who exported it to Persia, where it was used as dye.[343] The sticks of Lac prepared by the women of Surat were exported by English and the Dutch companies for about 150 chests annually. These sticks costed about 10 *sols* the liver.[344] Chief articles of trade from Satgaon were enormous quantity of silken fabrics, sometimes as many as 50 ship loads, large quantities of sugar, dried and conserved myrobalans, oranges and lemons. All the outgoing Cargo were not composed of goods from Bengal alone, but from all parts of India.[345] Under commission issued to Sir Henry Middleton and others for the 6th voyage to East Indies in 1610, it was laid down that *"the most desire able commodities for sending home are indigo… red sandal wood, 3 tons… and silk of Persia, a good quality."*[346] In 1777 the consignment of the articles exported from India by English company consisted of 4000 pieces of coconut[347] The manufacture of the perfumes of different kinds received a new stimulus from the foreigners. Persian were addicted to the articles-de-luxe manufactured commodities like perfumes, silk, etc. These commodities were sent to them on camels from the banks of Indus to those of the Oxus, down which river they were conveyed to Caspian, and thence circulated either by land-carriage, or by navigable rivers to the various parts of Persia.[348] Amongst the products exported from India sugar (in various forms) also finds mention. The profitability of the sugar trade is indicated by the fact that despite the greater weight of the sugar as compared to some other higher value goods, merchant transported long distances overland.[349] Barbosa considers sugar as the chief article of export from Bengal.[350] Sugar was frequently sent to Europe.[351] Moist sugar was exported in quantity from the kingdom of Bengal. Loaf sugar was produced at Ahmedabad, was also called royal sugar. These loaves of sugar generally weighed from 8 to 10 livers.[352] Manucci mentions the export of white and black sugar from the port of Lahari-Bandar in Sindh.[353] Brisk trade in sugar was carried from Punjab also. It appears that at least in the times of Pelsaert the north-western land route to Persia through Lahore was quite important for sugar trade. From Multan, sugar was sent down by the river to Thatta and also north to Lahore.[354] Bonford observed that sugar and sugar candy manufactured in Lahore were usually transported by Thatta merchants down the river and thence exported to Muscat, Congo, Bussara. Furthermore in 1640, Philip Wilde, on his way from Agra to Sindh bought sugar candy at Lahore.[355]

In the ship-building India appeared to have taken the lead. Writing in the 15th century, Conti has recorded the manufacture of ships of 1,000 tons in India, much larger than any with which he was familiar in the Mediterranean[356] Arabs used ships built in India for their trade in the Indian ocean.[357] Shipping remained an important industry up to the end of 18th century. The importance of the industry depended on the maintenance of an adequate supply of small ships for the sea going trade, and small boats for moving goods along the coast.[358] Many English merchants and others had their ships and vessels yearly built in India.[359] Popularity of the Indian ships can be estimated from the point that the ship builders of the port of London raised such a storm of protest that the court of Directors forbade the use of Indian ships and sailors. England had to ban Indian built ships from carrying goods to her ports to save its own ship building industry from competition.[360] The manufacture of Botanical products had widened the markets of India and had raised the prices of Indian commodities by increasing the aggregated demand.

There were botanical products which found a small market in India, these were raw-silk, fruits both fresh and dry, Persian perfumes, sandal wood, etc. An important consideration in this period was the busy overland caravan trade. The overland route via Kabul through the Khyber pass and Peshawar fed the markets of Lahore with fruits from Central Asia and silks and porcelain from China.[361] Fruits were the main commodities which were imported from Persia and Central Asia via Multan and Kabul. Both fresh and dry fruits such melons, peaches, pears, apples, almonds, quinces, pomegranates, dates, raisins, figs, plums, etc. were imported from Farghana, Bukhara, Badakshan and Samarqand.[362] Pelsaert mentions the import of fruits from Persia.[363] Bernier writes that Fruit markets at Delhi were well stocked with dry and fresh fruits from Persia, Balkh, Bokhara and Samarqand. Fresh grapes black and white, were imported, wrapped in cotton. Pears and apples of 3 or 4 sorts and *"those admire able melons (sarda) which last the whole winter were available. These fruits were, however, very dear; a single melon selling for a crown and a half."*[364] Marcopolo describes the export of dry fruits from Persia, specially dates and pistachios.[365] Some imported goods found in the markets of Benaras are mentioned in the price-list fixed by W.Hastings in 1781 are preserved in the papers relating to India, such as spices, saffron, dry fruits, etc. Among the dry fruits almonds, dates, raisins and *munaqqas* (large raisins) are included. They were sold at rupees 35, 16, 50 and 30 a *maund* respectively.[366] Another article which finds mention is silk. Silk from Persia was one of the commodities imported in to North India. Some came through Punjab and some by the way of Sindh. Merchants from Isfahan who came to Thatta to purchase Indian commodities brought silk for sale, even though its export from Persia was illegal.[367] The demand for silk in India was subject to fluctuations. It is learnt that around 1618, these merchants found little profit in carrying plain silk and preferred to *"carry the proceeds of the sold goods in ready money"*, or in *"sylke stuffs"*. Earlier they used to transport silk cloth and have it 'wrought into stuffs at Lahore.'[368] Silk from china was also imported by India.[369] Among the other articles which were imported by India were Sandal wood. Pelsaert writes:

Sandal wood is brought to Agra in moderate quantity from Portuguese who obtain it in Timor, and transport it to Malacca, whence it is carried to Goa and Cambay. No great trade can be done; 80 maunds or 400lb. may sell at not more than 50 rupees the maund.[370]

Perfumes, specially the *attar* of roses from Persia, was in great demand.[371] There was a great demand of Damishqi rose water also.[372] In the reign of Sultan Bahadur Shah of Gujarat, it was considered to be important import item. Among the various articles imported at the port of Kambayet, there was one thousand and 600 *maunds* of rose water from Damascus.[373] Among other articles that were imported by India were coconuts from Maldives islands;[374] citric acid flakes (made out of the juices of lemon, oranges and other such fruits after boiling the juice), *anardana*, grapes from Baihraich;[375] Goa used to send to Camby the wines, fruits, etc which it imported from Portugal.[376]

With the development of urban centres the demand for the Botanical products increased. These were related with almost every important economic activity of 16th and 17th centuries, whether it was manufacture of the means of the transport, material for the construction of houses, maintenance of the silk industry and manufacture of all other articles, people were dependent on the trees. Trees not only fulfilled the needs of the people, but also provided employment to various artisans as they were engaged in the manufacture of the various items, and these products were appreciated not only in the country, but in the foreign lands also. There was a brisk trade of these articles. All this widened the scope of Indian goods in the foreign markets and helped to improve the economy of the Empire.

References And Notes

1. *Baburnama*, pp-503-13; *Ain* Vol.1 pp-68-78.
2. Naqvi. *Urbanizations and Urban Centers*, p-46.
3. *Baburnama*, pp-503-513.
4. *Ain* Vol.1, p-70.
5. *Ibid*, p-70.
6. *Ibid*, p-69.
7. *India as seen by William Finch*, p-79.
8. Ellison. *Nurjahan*, p-247.
9. *Ain*, Vol.1, pp-70, 71.
10. Naqvi. *Urbanisations and Urban Centres*, p-24.
11. *Tuzuk* Vol.1, p-5.
12. Bernier, p -249.
13. *Geographical Glimpses*, Vol-3; *Khulasat*, p-62.
14. Tavernier Vol.1, pp-48, 65, 150.
15. Irfan Habib. *The Agrarian System*; p-55.
16. Qureshi. *The Administration of the Mughal Empire*, p-174.
17. Della Valle Vol.1, p-134.
18. The *Jahangirnama*, p-333.
19. *India under British Empire*, pp-111, 112.
20. Pelsaert, p-48.
21. *Ain* Vol.1, p-159.
22. *Baburnama*, p-505.
23. *Ain* Vol-1, p-237.
24. *Ibid*, p-237-239.
25. *Ibid*, p-239.
26. *Ibid*, pp-237-239.
27. *Ibid*, p-234.
28. *Ibid*, p-234.
29. *Ibid*, p-234.
30. *Ibid*, p-234.
31. *Tuzuk* Vol.1, pp-6, 7.
32. Naqvi. *Urbanization and Urban Centers*, p-34.
33. *Ibid*, p-34.
34. *India under British Empire*, pp-96, 98.
35. K.M.Ashraf. *Life and conditions*, p-120.
36. *Babur-Nama*, pp-514, 515.
37. *Ain* Vol.1, p-81.
38. *Ibid*, p-82.
39. *Ibid*, p-93.

40. *Tuzuk* Vol.I, p-5, 6.
41. *The Jahangirnama*, p-332.
42. *India as seen by William Finch*, p-79.
43. *Khulasat in Geographical Glimpses* Vol.3, p-63.
44. *Masalik-ul-Absar-Fi-Mumalik-ul-Ansar* by Shahabuddin-Al-Umar in *Geographical Glimpses*, Vol.1, p-31.
45. Ray, S.C. *History of Mughal India* (Comprehensive History of India vol.4) (from 1526-1707 AD) 2nd edition. New Delhi: People's Publishing House, 1990 p-329.
46. *Ibid*, p-329.
47. Bernier, p-44.
48. Tavernier, Vol.2, p-16.
49. Moosvi, Shirin. *Man and Nature*, symposium paper. published by IHC, 1993 pp-18-19.
50. Sarkar, Jagdish Narayan. *Mughal Economy: Organisation and Working*. Calcutta: Naya Prakash, 1987 P-58.
51. *Ain*. Vol.1, p-87.
52. *Ibid*, p-87.
53. *Ibid*, pp-83-84.
54. *Ibid*, p-87.
55. *Ain*. Vol.1, p-93.
56. *Ain* Vol.1, pp-352, 353.
57. *Ibid*, p-352, 353; Also see *Khulasat*, p-112.
58. Bernier, p-398.
59. Pelsaert, p-33.
60. *The Jahangirnama*, p-332.
61. *Ibid*, p-244.
62. *Ain* Vol.1, p-233.
63. *Baburnama*, p-505; also see *Ain* Vol.1 p-75 and Zain Khan's *Tabqat-i-Baburi* p-120.
64. *Ain* Vol.1, p-92.
65. *Ibid*, p-231.
66. Bernier, p-248.
67. *Ibid*, p-248.
68. *Mirat-i-Ahamdi of* Mohammad Khan in Geographical Glimpses, Vol.3, p-39.
69. Father Monserrate, p-103.
70. *Ibid*, pp-104, 108, 110.
71. Farooque. *Roads and Communications*; p-51.
72. Bernier, p-398.
73. *The Jahangirnama*, pp 324, 325.

74. *Ibid*, p-331.
75. *The Shahjahan-nama*, p-125.
76. Farooque. *Roads and Communications*, p-52.
77. Mackenzie, Gordon. *A Manual of Kistna District in the Presidency of Madras.* Compiled for the Government of Madras. Madras: 1883 p-167.
78. Father Monserrate, p-81.
79. *India Under British Empire*, p-97.
80. *Ibid*, p-97.
81. *The Shahjahan-nama*, p-125.
82. Naqvi. *Urbanisation and Urban Centres*, p-46.
83. *Ain* Vol.1, pp-237-239.
84. Naqvi. *op.cit.*, p-46.
85. *Ibid*, p-45.
86. *Ibid*. p-45. Also See *India at the Death of Akbar*; pp-135, 157.
87. Naqvi, *op.cit.*, p-45. Also see Chetan Singh, p-182.
88. Singh, Chetan. *Region and Empire, Punjab in the 17th Century.* New Delhi: Oxford University Press, 1991 p-186. Also see Muzzafar Alam, p-201.
89. *Ibid*, pp-188, 200.
90. Shirin Moosvi. *Man and Nature*, p-1.
91. *India at the Death of Akbar*; -135.
92. *Ain*, Vol.1, p-290.
93. *Ain* Vol.2, p-339.
94. *The Jahangirnama*, p-331.
95. *Ain*, Vol.1, p-290.
96. Naqvi. *op.cit.*, p-34. Also *see India at the Death of Akbar*, p-135.
97. *Ibid*, p-45.
98. *Ibid*, p-45. Also See Chetan Singh, p-211.
99. Chetan Singh. *Region and Empire*, p-211.
100. *Khulasat*; p-77.
101. Chetan Singh. *Region and Empire*, p-212.
102. *Ibid*, p-212.
103. Naqvi *Urbanisation and Urban Centres* p-46.
103a. Parmu, R.K. *A History of Muslim Rule in Kashmir* 1320-1819. Srinagar: Gulshan Books, 2009 pp-416,417
104. *India under the British Empire* p-96.
105. Qureshi.*The Administration of the Mughal Empire*, pp-175, 256, 257.
106. *Early Travels*, pp-154, 278, 329, 331.
107. *Mirat-i-Ahmadi* in Geographical Glimpses, Vol.3, p-39.
108. *Ain* Vol.1, p-72. Also see Manucci, Vol.3, p-171.
109. *Ain* Vol.1, p-76.
110. Manucci Vol.3; p-176.

111. *Ibid,* p-176.
112. *Ibid,* p-177.
113. *Ibid,* pl-73.
114. *Ibid,* p-173.
115. *Ibid,* p-171.
116. Zain Khan's *Tabqat-i-Baburi* p-122.
117. Pelsaert, p-65.
118. P.N. Chopra. *Some aspects of Society and Culture during Mughal Age* p-18.
119. *The Jahangirnama,* p-184.
120. *India under British Empire,* p-117.
121. *Ibid,* p-119.
122. *Baburnama,* p-511.
123. Careri, J.F.Gemelli. *A Voyage round the World by John Francis Gemelli, Careri.* ed. by Surendranath Sen. New Delhi: National Archives of India, 1949 Part 3; p-202.
124. Bernier, p-389. Also see *Aurangzeb in Kashmir*; p-62.
125. Manucci Vol.1, p-46.
126. *India under British Empire,* p-120.
127. *Ibid,* pp-121, 124.
128. *Ibid,* p-125.
128a. Edward Terry, p-98
129. *Ibid,* pp-98, 99.
130. Manucci Vol.3, p-176. Also see *Ain* Vol.1, p-76.
131. *India under British Empire,* p-99.
132. *Ain* Vol.1, p-76.
133. *Baburnama,* p-508. Also see Zain Khan's *Tabqat-i-Baburi,* p-122.
134. *Baburnama,* p-508. Also see Ain Vol.1, p-76 and Zain Khan, p-122.
135. Manucci Vol.3, p-176.
136. *India under British Empire,* p-99.
137. Manucci Vol.3, pl-75.
138. *Ibid* p-177.
139. Maheshwar Dayal. *Rediscovering Delhi,* p-36.
140. *Early Travels,* p-13.
141. Della Valle Vol.2, pp-294, 295, 327.
142. *Ibid,* p-327. *Also see India under British Empire,* p-120.
143. Zain Khan's *Tabqat-i-Baburi,* pp-122, 123.
144. *Ain* vol.1, p-92.
145. Ghulam Hussain Zaidpuri's *Riyaz-us-Salatin* in *Geographical Glimpses,* Vol.3, p-58.
146. *India under British Empire,* p-98.
147. *Ibid,* pp-98, 119, 114, 115, 113.

148. K.M. Ashraf. *Life and Conditions*, p-133.
149. *India under British Empire*, p-96.
150. Farouki. *Aurangzeb and His Times*, p-494.
151. Bernier, p-402.
151a. R.K.Parmu, pp-416,417
152. De Laet, p-8.
152a Edward Terry, p-128
153. Naqvi. *Urbanisation and Urban Centers*, p-34.
154. *India at the Death of Akbar*, p-255.
155. *Baburnama*, p-510. Also see Zain Khan, p-122, 123.
156. *Khulasat*, p-112. Also see *Ain* Vol.2, p-354.
157. *Ain Vol.2, p-138. Amal-i-Saleh of Mohammad Saleh Kamboh in Geographical Glimpses, Vol.3; p-13.*
158. *India at the Death of Akbar*, p-153.
159. *India under British Empire*, p-98.
160. *Khulasat in Geographical Glimpses, Vol.3, p-73.*
161. K.M. Ashraf. *Life and Conditions*, p-133.
162. *India at the Death of Akbar*, p-149.
163. *Baburnama*, p-505. Also see Zain Khan, p-120 and *Ain* Vol.1, p-75.
164. Nanda, Meera. *European Travel Accounts during the Reigns of Shahjahan and Aurangzeb*. Kurukshetra: Nirmal Book Agency, 1994 p-96.
165. *Ain*, Vol.1, p-75.
166. De Laet, p-24.
167. *Baburnama*, p-509. Also see Zain Khan, p-122, 123.
168. Manucci Vol.3, p-176; Careri. *A Voyage Round the World* part-3, p-200.
169. *Baburnama*, p-509.
170. *Ibid*, p-509.
171. *Ibid*, p-509.
172. Tavernier vol.1, p-128.
173. *Early Travels*, pp-13, 175, 297, 298, 300.
174. Careri. *A Voyage*, p-200.
175. Manucci, Vol.3, p-177.
176. P.N. Chopra. *Some Aspects of Social Life During Mughal Age*, p-48.
177. Ain Vol1, p-73.
178. P.N. Chopra. *op.cit.*, p-48.
179. D.Pant. *The Commercial Policy of the Mughals*, p-19.
180. *Ibid.*, p-19, 20.
181. Naqvi, H.K. *Dyeing of Cotton goods in Hindustan* (1556-1803). Ahmedabad: The Journal of Indian Textile Industry. No.VII, 1967 pp-45, 56.
182. Naqvi. *Urbanization and Urban Centres*, p –45.
183. *Ain* Vol.1, p-92.

184. *Ibid.*, p-92.
185. Irfan Habib. *The Agrarian system*, p-58.
186. Tavernier Vol.2, p-18.
187. *Ibid.*, p-221.
188. *India under British Empire*, p-117.
189. *Ibid*; p-124.
190. Naqvi; *Urbanisation and Urban Centers*; pp-140, 141.
191. *Ibid*; p-40.
192. Tavernier Vol.2, p-2; *Ain* Vol.2, p-353.
193. *The Tarikh-i-Rashidi*, p-425.
194. *Ain* Vol.2, p-353. Also see *Khulasat*, p-121.
195. *Tarikh-i-Rashidi*, p-425. *Ain* Vol.2, p-353.
196. *Tarikh-i-Rashidi*, p-425.
197. *The Akbarnama*, Vol.3, p-725.
198. *Tuzuk* Vol.2, p-146.
199. Bernier, p-442.
200. Naqvi. *Urbanisation and Urban Centres*, p-47.
201. *Ain* Vol.2, p-136.
202. Naqvi. *Urbanisation and Urban Centres*, p-47.
203. Tavernier Vol.2, p-2.
204. *Ibid.*, p-2.
205. *Ibid.*, p-3.
206. Naqvi. *Urbanisation and Urban Centres*, p-47.
207. Bernier, p-439-40.
208. Irfan Habib. *The Agrarian System*, p-58.
209. Naqvi. *Urbanisation and Urban Centres*, p-46. Also see Irfan Habib; Agrarian System; p-58.
210. Carre, Abbe. *The travels of Abbe Carre*. Vol.3 New Delhi: Asian Educational Services,1990 p-326. References are probably of tassar.
211. *Early Travels, p-26*.
212. *India at the Death of Akbar*, p-164.
213. Irfan *Habib.The Agrarian System*, p-58.
214. Tavernier Vol.2, p-220.
215. *Ibid.*, p-220.
216. *Ain* Vol.1, pp-79-80.
217. *Ibid.*, p-81.
218. *Ibid.*, pp-80-82.
219. *Baburbama*, p-514.
220. P.N.Chopra. *Some aspects of Society and Culture During Mughal Age*, p-16.
221. K.M. Ashraf. *Life and Conditions*, p-121, 122.
222. *Ain* Vol.1, p-80.

223. P.N.Chopra. *op.cit.*, p-16.
224. *Journal of Pakistan Historical Society*, p-309.
225. *The Jahangirnama*, p-163.
226. Manucci Vol.1, pp-158, 159.
227. Naqvi. *Urbanisation and Urban Centres*, p-44.
228. *Ain* Vol.1, p-249.
229. De Laet, p-38.
230. Naqvi. *Urbanization and Urban Centres*, pp-44, 45.
231. *Ain* Vol.1, pp-80-87.
232. *Ibid.*, p-86.
233. *Ain* Vol.1, p-87.
234. *Ibid.*, p-79.
235. P.N.Chopra. *Some Aspects of social Life during the Mughal Age*, p-18.
236. Pelsaert, p-65.
237. P.N.Chopra. *op.cit.*, p-18.
238. Manucci Vol.3, pp-315-16.
239. S.C.Ray. *History of Mughal India*, p-329.
240. Tavernier Vol.2, p-18.
241. *Ibid.*, p-18.
242. *Ibid.*, p-19.
243. S.C.Ray. *History of Mughal India*, p-329.
244. D.Pant. *The Commercial Policy of Mughals*, p-97.
245. *Ain* Vol.1, p-89; Also see *Khulasat*, p-113.
246. D.Pant. *op.cit.*, p-97.
247. *Ain* Vol.1, pp-23-25.
248. *Khulasat*; p-119.
249. *Ain* Vol.1, p-234.
250. *Ibid.*, p-234.
251. *Ibid.*, p-234.
252. *Ibid.*, p-234.
253. *Ibid.*, p-234.
254. *Ibid.*, p-234.
255. *Ibid.*, p-234.
256. *India under British Empire*, p-116.
257. M.A.Ansari, *Geographical Glimpses*, Vol.3, Intr.[xxii].
258. *Ain* Vol.1, p-62.
259. Bernier, p-438.
260. *Ibid.*, p-438.
261. Naqvi, H.K. *Urban centres and Industries in Upper India* (1556 – 1803). Bombay: Asia Publishing House, 1968 pp-52-53.
262. Manucci Vol.3, p-171.

263. *Ain* Vol.1, p-67.
264. *Ibid.*, p-274.
265. *Ibid.*, pp-236, 237.
266. Shirin Moosvi. *Man and Nature*, p-18.
267. Irfan Habib. *The Agrarian System*, p-64.
268. K.M.Ashraf. *Life and Conditions*, p-134.
269. *India at the Death of Akbar*, p-147.
270. K.M.Ashraf. *op.cit.*, p-134.
271. Tavernier Vol.1, p-238.
272. Pelsaert, p-32, 31.
273. Chetan Singh. *Region and Empire*, p-108.
274. Manucci Vol.2, p-399.
275. Tavernier Vol.2, p-16.
276. *Ibid.*, p-222.
277. Naqvi. *Urbanization and Urban Centres*, pp-67-68.
278. Farooqe. *Roads and Communications*, p-55.
279. *Tuzuk* Vol.2, p-100.
280. Manucci Vol.1, p-159.
281. *India as seen by William Finch*, p-72.
282. K.M.Ashraf. *Life and Conditions*, p-125.
283. Qureshi. *The Administration of the Mughal Empire*, p-59.
284. Bernier, pp-258, 259.
285. K.M.Ashraf. *Life and Conditions*, p-125.
286. Naqvi. *Urbanization and Urban Centres*, p-47.
287. *India at the Death of Akbar*, p-164.
288. C.A.Bayley, p-144.
289. *Ibid.*, p-146.
290. *Ibid.*, p-147.
291. *Ain* Vol.1, p-68.
292. *Ibid.*, p-69.
293. *Ibid.*, pp-235, 236.
294. *Ibid.*, p-235.
295. *Ibid.*, p-235.
296. *Ibid.*, p-235.
297. *Ibid.*, p-236.
298. *India at the Death of Akbar*, p-157.
299. *Ain* vol.1, p-290; *Ain* Vol.2, p-354.
300. *The Akbarnama* Vol.3, p-995.
301. *Ibid.*, p-1001.
302. Chetan Singh. *Region and Empire*, p-184.
303. Bernier, p-402.

304. Chetan Singh. *Region and Empire*, p-184.
305. K.M.Ashraf. *Life and Conditions*, p-121.
306. Tavernier Vol.2, p-19.
307. Tavernier Vol.1, p-128.
308. Irfan Habib. *Agrarian System*, p-65.
309. Qureshi. *The Administration of the Mughal Empire*, p-57.
310. *Khulasat*, p-112.
311. Naqvi. *Urbanisation and Urban Centres*; p-47.
312. Qureshi. *The Administration of the Mughal Empire*, pp-61, 62.
313. Nizami, K.A. *State and Culture in Medieval India*. New Delhi: Adam Publishers,1985 p-250.
314. Srivastava, M.P. *Policies of the Great Mughals*. Allahabad: Chugh Publications, 1978 p-51.
315. Naqvi. *Urbanization and Urban Centres*, pp-110-111.
316. *Ain* Vol.1, pp-89-90. Also see *Tuzuk* Vol.1, p-92.
317. *Tuzuk* Vol.2, pp-146, 159.
318. Manucci Vol.2, p-428.
319. *Tuzuk* Vol.2, p-146.
320. K.M.Ashraf. *Life and Conditions*, p-145.
321. Tavernier Vol.2, pp-2-3.
322. D.Pant. *The Commercial Policy of Mughals*, p-238.
323. *Ibid.*, pp-92, 93.
324. *The Journal of Pakistan Historical Society*, p-309.
325. M.P.Srivasatava. *op.cit.*, pp-43, 50.
326. Faruki. *Aurangzeb and His Times*, pp-494, 495.
327. *Ibid.*, p-494.
328. D.Pant. *The Commercial Policy of the Mughals*, p-97.
329. Pelsaert, pp-35, 36.
330. *Ain* Vol.2, p-352.
331. Naqvi. *Urbanisation and Urban Centres*, p-46.
332. *Early Travels*, p-169.
333. *Khulasat*, p-106.
334. Pelsaert, p-19.
335. Irfan Habib. *Agrarian System*, p-78.
336. *Ibid.*, pp-78, 79.
337. *Early Travels*, p-44.
338. D.Pant. *op.cit.*, p-141.
339. Tavernier Vol.2, p-2.
340. Ibid., p-3.
341. Irfan Habib. *Agrarian System*, p-81.
342. Naqvi. *Urbanisation and Urban Centres*, p-98.

343. Tavernier Vol.2, p-18.
344. *Ibid.*, p-19.
345. Naqvi. *Urabanization and Urban Centres*, p-136.
346. D. Pant. *The Commercial Policy of the Mughals*, p-157.
347. Naqvi. *Upper India [1556 – 1803]*, p-110.
348. D.Pant. *op.cit.* p-8.
349. Chetan Singh. *Region and Empire*, p-218.
350. K.M.Ashraf. *Life and Conditions*, p-145.
351. Irfan Habib. *Agrarian System*, p-81.
352. Tavernier Vol.2, p-19.
353. Manucci Vol.1, p-59.
354. Pelsaert, p-31.
355. Chetan Singh. *op.cit.*, p-218.
356. *India at the Death of Akbar*, pp-159-160.
357. Qureshi. *The Administration of Mughal Empire*, p-256.
358. *India at the Death of Akbar*, p-160.
359. Qureshi. *op.cit.*, p-256.
360. *Ibid.*, p-257.
361. John F.Richard, *The New Cambridge History of India*, p-50.
362. *Ain* Vol.1, p-68-69.
363. Pelsaert, p-31.
364. Bernier, p-249.
365. D.Pant. *The Commercial Policy of the Mughals*, p-11.
366. Naqvi. *Upper India [1556 – 1803]* pp-96, 97.
367. Pelsaert, p-32.
368. Chetan Singh. *Region and Empire*, p-222.
369. D.Pant. *op.cit.*, p-106.
370. Pelsaert, p-25.
371. D.Pant. *op.cit.*, p-106.
372. Naqvi. *Urbanisation and Urban Centres*, p-111.
373. Nizamuddin Ahmad. *Tabqat-i-Akbari*, Vol.1, p-
374. Naqvi. *Urbanization and Urban Centres*, p-110.
375. *Khulasat in Geographical Glimpses* Vol.III, p-72.
376. D.Pant. *The Commercial Policy of the Mughals*, p-154.

PRESERVATION OF FRUITS AND FLOWERS

With the progress in horticulture and the process of its intensification, the industry of fruits and flowers preservation became one of the dominant features of the economy of the Mughal empire. Although references to the fruits and flowers preservation in forms of pickles, scents and perfumes[1] are found in ancient India. It was during the Mughal period that this industry received impetus. Fruits and flowers were preserved in different forms such as pickles, marmalades, *murabbas*, scents and perfumes. Also the process of import and export of the fruits was accelerated. Consequently different methods were employed to keep them fresh and to avoid their spoilage.[2] Abul Fazl writes that various fruits which were preserved in form of pickles, were lemons, sugarcane, apples, bamboos, quinces, raisins, *munaqqa*, peaches, etc.[3] Citrons were also preserved.[4] Pomegranates and various citric fruits such as oranges were also preserved by the people.[5] Odoriferous plants of both foreign and Indian origin were used for extracting scented oils and perfumes.[6] But the more sophisticated perfumes were extracted from the flowers having stronger or more refined scents such as that of roses, *gul-i-henna*, jasmine, saffron, *sewti*, *chambeli*, ray-bel, *ketki*, *mongra*, *champa*, *kuza*, *padal*, *juhi*, *niwari*, *kewara*, *chalta*, *gulal*, *tasbih gulal*, *singarhar*, violet, *karna*, *kapur-bel*, *gul-i-zafaran*, etc.[7] Besides, various types of tree products such as sandal (*chandan*), aloe wood, etc. were used in the manufacturing of the perfumes.[8] Sujan Rai has praised the flowers of India especially *Kewra* and *Ketki*. He writes:

However, Kewra and ketki are exceptional in delicacy of their smell and refreshness. So much so that a hand full of these can provide beautiful aroma for a gathering of hundred persons. It fills not only the house but *the whole locality with aroma. This smell last long. These flowers are not to be found in Iran or Turan*⁹

Various techniques for the conservation of fruits were used under the Mughals. Different methods were used for the preservation of mangoes. For instance, Daulat Khan Lodi sent Babur a gift of mangoes preserved in honey[10]. About the preservation of mangoes, Babur writes mangoes, when *unripe, they make excellent condiments (qatiq), are good also when preserved in syrup.*[11]

Abul Fazl writes that mangoes were preserved by using the warm wax and honey,

If a half ripe mango, together with its stalk to a length of about two fingers, be taken from the tree, and the broken end of its stalk be closed with warm wax, and kept in butter, or honey, the fruit will retain its taste for two or three months, whilst the colour will remain even for a year[12].

Manucci also mentions about the preservation of mangoes:

They also make the fruits into preserves, which are exported to various places...[13]

Jahangir has praised one of his nobles Muqarrab Khan for the efforts that he made for the preservation of mangoes:

Mangoes are not found in Hindustan past the end of the month of Tir (June-July), but Muqarrab Khan has made orchards in the pargana of Kairana, his ancestral home land, in which they can be kept and preserved some how for an extra two months after their season so that mangoes were sent to the fruit store house in Ajmer every day.[14]

Fruits were preserved by various other methods also. Terry writes that fruits were preserved in sugar.[15] Bernier writes that in Bengal people preserved citrons in large quantities and he writes that method employed for the preservation of citrons by the people of Bengal was similar to that of Europe[16] Babur writes that citrons were preserved in the form of *sherbet*. He writes:

... the other (citron), that of Hindustan and Bajaur, is acid, quite deliciously acid, and makes excellent sherbet, well flavoured and wholesome drinking.[17]

He also writes that from the juice of *sangatra* also was made *"a very pleasant and whole some sherbet"*.[18] Della Valle has mentioned about the preservation of myrobalans in sugar.[19] He writes that generally myrobalans did not have pleasant taste "but when preserved becomes good."[20] People used to make confection (*murabba*) out of *amla* because of its medicinal values.[21] Manucci found the preservation of the palm fruit (*palmeria brava*) in form of marmalade. He writes:

After the fruit has ripened the pulp hardens, and the juice is drawn from the shell; leaving this for a few days in a sun, it becomes a sort of marmalade, eaten by many of the natives in place of bread.[22]

Seeds of the pomegranates were dried in sun and were preserved in form of *anardana*.[23] People used citric acid flakes made of citric fruits. These were *made out of the juices of lemon, oranges and other such fruits after boiling their juice*[24]

During this period, the loads of fruits were being imported from Kabul, Qandhar and Kashmir.[25] These fruits were packed with so much care that they looked fresh when brought to Indian markets. As is obvious from the Jahangir's memoirs:

At this place some melons came from Kariz, which is a town dependent on Herat, and it is certain that in Khurasan there are no melons better than those of Kariz. Although it is at a distance of 1,400 kos, and kafilahs (carvans) take five months to come, they arrived very ripe and fresh… together with these there came oranges (kaunla) from Bengal, and though that place is 1,000 kos distant most of them arrived quite fresh.[26]

Bernier writes that in winter *"excellent fresh grapes black and white"* were brought from Persia, Balkh, Bokara and Samarqand, by wrapping them in cotton. The grapes were embedded in cotton wool in the round wooden boxes which kept them fresh for months.[27] The *sahibi*, a grape praised by Babur amongst Samarqandi fruits, which grew in Koh-i-daman; another well known grape of Kabul was the long stone less *husaini*, these were brought by Afghan traders into Hindustan in round, flat boxes of poplar wood.[28] Flowers were also preserved in many ways. Flowers of *mahuwa* tree were dried and eaten like raisins.[29] A flower named *Tasbih gulal* had a very fine smell. People used to make rosaries of the flowers, which kept fresh for a week.[30] The *padal* flower gave the water an agreeable flavour and smell. It was on this account that people preserved the flowers, mixed with clay, for such times when the flower was out of season.[31]

Preservation of fruits in form of pickles was the most popular method. Abul Fazl gives the list of different types of pickles for example of mangoes, sour limes, lemons, apples, raisin, and *karil* buds.[32] Mangoes were pickled in oil and vinegar.[33] The lemons were preserved in oil, vinegar, salt and lemon juice.[34] Abul Fazl has also mentioned that pickled bamboos, pickled apples, quinces, raisins and *munaqqa* and peaches were available in the market.[35] People also made the pickle of *Karil* buds (flower).[36] Manucci mentions about various kinds of mango pickles, which could be kept in good conditions even for two years[37].

Different types perfumes and scented oils were extracted from the scented flowers and scented woods. Different varieties of *itar* (scents) and oils were in use, especially by the high class and aristocracy. Abul Fazl writes:

His Majesty is very fond of perfumes and encourages this department from the religious motives. The court hall is continually scented with ambergris, aloewood, and compositions according to ancient recepies, or mixtures invented by His Majesty; and the incense is daily burnt in gold and silver censers of various shapes; whilst sweet-smelling flowers were used in large quantities. Oil are also extracted from flowers, and used for skin and hair.[38]

Abul Fazl has given a long list of the scented oils and incenses and the methods used for their preparation. *Santuk* was used for keeping the skin fresh and it was prepared by mixing civet, *chuwa*, *chambeli* essence, and rose water.[39] *Argaja* was used in summer for keeping the skin cool; it was made by mixing sandal wood, *chuwa*, violet root, camphor and eleven, bottles of rose water,[40] for the preparation of *Gulkama* were required ambergris, ladan, best musk, wood of aloes, all these were put into porcelain vessel and mixed with the juice of a flower called *gul-i-surkh* and then it was exposed to sun till it dried up and in the evening it was wetted with rose water and with the extract of a flower called *bahar* and was pounded on *samaq* stone. Then it was allowed to stand for ten days, and mixed with the juice of a flower called *bahar-i-naranj* (orange flower bloom) and then again it was kept to dry. During next twenty days was occasionally added the mixture of black *rayhan* (sweet basil).[41] *Ruh-afza* was made by mixing aloe wood, sandal wood, ladan, rose water, etc. It was burnt in censers, and smelled very fine.[42] *Opatna* was the scented soap; it was made by boiling the mixture of *ladan*, aloe wood, *bahar-i-naranj* (orange flower blossom), sandal wood, apples, rose water and extract of bahar, etc.[43] *Abirmaya* was prepared by boiling sandal wood; violet root, *ladan*, *bahar-i-naranj* in about ten bottles of rose water. *Kishta* was mixture of aloe wood, *ladan*, sandal wood, musk, rose water. It was made into discs. It smelled very fine when burnt, and was exhilarating. *Bukhur* was the mixture of aloe wood, sandal wood, *ladan*, sugar, rose water, etc.[44] The other perfumes were *fatila, Bajrat, abirlkisr* and *Ghasul. Ghasul* was the liquid soap which was made by mixing sandal wood, camphor, etc. with rose water.[45] Manrique has referred to the preparation of the rich scented oils at Mickapur (in Bengal) which were made of various odoriferous flowers and other scented ingredients and with which men and women anointed themselves while taking their accustomed daily baths.[46] Oil was also extracted from the seeds of *mahuwa* flower.[47] Jahangir writes that from *ketki, keora* and *chambeli*; sweet scented oils were extracted.[48] Manucci refers to the extraction of oil from coconut.[49] Manucci also referred to the preparation of jasmine oil in Gwalior.[50]

Royal ladies took keen interest in inventing various kinds of *itar*. Pelsaert noted that women studied

night and day how to make exciting perfumes and efficacious preserves, such as mosseri or falroj containing amber, pearls, gold, amboa, opium, and other stimulants[51]

We know from Abul Fazl that the imperial kitchen was elaborately equipped for all kinds of techniques that might be needed in the process of drying, extracting, fermenting, distilling and straining. The *Ain-i-Akbari* provides a list of listed thirty four different types of perfumes such as *amber-i-ashhab*, musk, lignum aloes, *chuwa* (distilled wood of aloes), *Gaura*, *Bhimsini* camphor, *zafaran* from Kashmir, sandal wood, *kalanbak*, rose water, violet root, *sugandh gugala*, *alak*, etc.[52] One of the most famous perfume which was discovered during the reign of Jahangir was *itr-i-Jahangiri*. There is a story in Jahangir's Memoirs, which attributes the discovery of this *attar* of roses to Nurjahan's mother, Asmat Begam,

"when she was making rose water", once, Jahangir noted, *"a scum formed on the surface of the dishes into which the hot rose water was poured from jugs. She collected this scum little by little; and discovered that it was so strong "that if one drop be rubbed on the palm of the hand it scents a whole assembly, and it appears as if many many red rose buds had bloomed at once. It was such a good perfume that Jahangir "presented a string of pearls to the inventress," and Salima Sultan Begam gave the oil the name "Jahangir itr"*[53]

But this story was falsely attributed to Nurjahan by Manucci, who writes that in the after math of a scrap between husband and wife, Nurjahan decided to please Jahangir by giving him a large banquet. She filled all the reservoirs in the palace and gardens with roses in water and prohibited any one from washing hands in them. She happened, however, to fall asleep by one of the tanks and when she awoke she notice that a film of oil lay on the top of the water.

Furious at the thought that "some one had thrown fat into this tank," she had the oil tested on the finger tips of a companion. Finding that it smelled very sweet and that it must have come directly from the rose petals, Nurjahan rubbed some all over her clothes and ran to awake the king who became *"lost in admiration at such a fine perfume." "It was thus"* concluded Manucci," that the secret of essence of roses was discovered in Hindustan"[54] Nurjahan Begam is also credited with the invention of extraction of inexpensive scent for common people, since *Itr-i-Jahangiri* was very costly.[55] Perfume was also extracted from the Kewra flower. Babur writes:

In amongst these inner leaves grow thing like what belongs to the middle of a flower, and from these things comes the excellent perfume.[56]

In *Ain* recipes have been given for the preparation of different perfumes such as amber, *ladan*, camphor, *zabad*, etc.[57] For the preparation of the perfume, *zabad* was first washed with rose water.

They then smear the zabad on the inside of the cup, keep it at night inverted in extract of chambeli, or Ray-bel, or surkh gul, or Gul-i-karna and expose it at day time to the rays of sun, covered with a piece of white cloth till all the moisture goes away. It may then be used, mixed with little rose-water.[58]

Aloe wood was often used in compound perfumes; when eaten it was exhilarating. It was generally employed in incense; the better qualities in form of a powder were often used for rubbing into skin and clothes.[59] Sandal wood was pounded and rubbed over the skin, but it was used in many other ways also.[60] *Kalanbak* (calembic) was the wood of a tree, it was pounded and was used for compound perfumes and people also made rosaries of it. *Sugandha gugala* (bdellium) was a plant very common in Hindustan; it was also used in the preparation of perfumes.[61] About the preparation of the camphor; Tavernier writes:

The cinnamon tree bears a fruit like an olive, but it not eatable. The Portuguese used together quantities of it, which they placed in cauldrons with water together with the small points of the ends of the branches, and they boiled the whole till the water was evaporated; when cooled the upper portion of what remained was like a paste of white wax, and at the bottom of the cauldron there was camphor. Of this paste they made tapers, which they used in churches during the services at the annual festivals and as soon as the tapers were lighted the church was perfumed throughout with an odour of cinnamon. They have often been sent to Lisbon for king's chapel.[62]

There were many fruits and flowers from which different types intoxicants were extracted. Babur writes that the spirit (*araq*) was extracted from both the fresh and dried flowers of *mahuwa*.[63] Abul Fzal too writes that from the fruit of *mahuwa*, which was called *gilaunda*, an intoxicating liquor was yielded.[64] An intoxicating liquor was also made from the use sugarcane. Abul Fazl has described various ways of its preparation.

One way is to follow. They pound Babul bark mixing it at the rate of 10 sets to one man of sugarcane, and put three times as much water over it. Then they take large jars, fill them with the mixture, and put them into the ground, surrounding them with dry horse dung. From 7 to 10 days are required to produce fermentation. It is a sign of perfection, when it has a sweet but stringent taste. When the liquor is to be strong, they again put to the mixture some brown sugar, and some times even drugs and perfumes, as ambergris, camphor, etc...This beverage when strained, may be use, but is mostly employed for the preparation of arrack.[64a]

The date palm (*khurma*) yielded another intoxicant. If it was drunk fresh it was sweet and pleasant, but if it was kept for a day or more it was exhilarating.[65] Manucci calls this liquor as *sura*. He writes:

when they want to draw the liquid, it is first necessary to open the flower, it being then in its rind; the end is cut, and then they begin to depress the whole flower in order to collect the juice, and fixing it, they attach to the end of earthern cup into which the substance drips. This is done twice a day, and flower yields each times twenty ounces, more or less, of the distillation. When drunk fresh it is sweet and suave, but kept for 12 hours, it tastes like beer and goes to the head. It is used by Indian in place of wine. From this liquid are manufactured aqua ardent, vinegar and sugar.[66]

Tarkul or palm tree produces a juice called *tari* which could be extracted three times a day.[67] Abul Fazl writes:

The juice is called Tari; when fresh it is sweet; when it is allowed to stand for some time it turns sub-acid and is inebriating.[68]

Babur writes that people extracted the juice by hanging the pot on the tree, "*it is said to be more exhilarating than date liquor,*"[69] Tavernier writes that the liquid which was obtained from palms was called "*suri*".[70] De laet too informs that from the palm-tree

… the people obtained a certain liquor which they call Tarrien [Tari, toddy] and suren [sura, liquor].[71]

Wine was made from the grapes which grew in the mountain tracts of Kashmir.[72] Terry writes that some common quality of wine was made out of sugar and a spicy kind of tree, it was called "*jagra*", and people named it "*rak*"[73] About *toddy*, Terry writes that toddy was extracted from an "spungy tree". The people climbed the trees very fast, under their soft branches they hung pots made of large and light "*gourd*". The toddy was distilled in the night. It was taken out before the sun shone on it. It was "pleasing to the taste as any new wine, very piercing, medicinal and inoffensive drink. "when left in sun it became strong. It was good for stone as it relieved its pain and was cheap too.[74] Liquor was also received from the coco tree. Careri says:

It it so nourishing that Indians live upon it without any other sustenance.[75]

Apart from these liquors there were different types of sherbets made from the juices of various fruits. For instance, rose water and sugar was added to the juice of lime, pomegranates, and the like to make the *sherbets*.[76]

The preserved products were in great demand. Contemporary Persian and vernacular sources as well as travel accounts reveal that Indians were fond of using a great variety of pickles *(achars)*, relishes *(chutneys)* and flavours of different kinds in order to stimulate the tongue and to whip up the action of stomach.[77] Both the middle and upper classes had a special taste for *achars* of mangoes and various conserves of water-melons, grapes, lemons, and oranges.[78] These pickles also formed the part of the diet of the Europeans who settled in India during this time. For instance, there is a reference to the use of mango pickle by Britishers at Surat[79]. In Chanaram Chakravarty's *"Dharam Mangal"*, there is a reference to the preparation of a very rich food and it contained mango pickle and relishes along with different sorts of delicious dishes.[80] Various types of *sherbets* were taken by the people. Rose water, sherbet and lemon juice mixed with ice were taken by the rich[81].

Sweet scented oils of various kinds were in great demand, people applied them to the hair and also rubbed on the body. How very essential oil was for a bath is clear from the words of Mukundram, a poet of 6th century when he writes *"My bath was without oil, water only was my drink and food and my infant child cried for hunger."*[82] Poor people used coconut oil and nobles would anoint their bodies with sandal and other oils extracted from various flowers.[83] In Gujarat, according to Barbosa, people anointed themselves with white sandal wood paste mixed with saffron and other scents.[84] In the hot weather the rich would add rose water to their tubs to keep their skin cool.[85] Perfumes like *Santak* and *Argajah* were also in great demand.[86] People used a sweat powder like that of sandal wood to get sweat out of their bodies and head and daubed the head with oil.[87] There were bathing houses *(hamams)* in almost all the important cities of the Mughal empire. In Agra alone, for example there were about 800 such houses doing a flourishing business. Adequate bathing arrangements were made at these places where various kinds of perfumed oils, and essences of sandals, cloves and oranges were freely supplied to the customers.[88] In the royal palaces and the mansions of aristocrats there were hot water baths or "*hamams*", which were used for bathing purposes during winter. Rose water with a variety of scents *(itr)*, perfumed oils and ointments were used by the well to do classes while taking bath. The Hindus of higher classes applied musk sandal wood paste, incense (*dhup*) and other varieties of indigenous perfumes to their bodies after their usual baths.[89] Perfumes were in great demand especially among the upper classes. Precious scents of diverse kinds were in use. Kaultilay's *Arthashastra* gives a long list of fragrant substances for the toilet preparations.[90] Akbar was extremely fond of perfumes and his private chamber was constantly scented with flowers. Incenses were always burned in all his rooms in gold and silver vessels.[91] Jahangir also promoted the culture of perfumes. *Itr-i-jahangiri* which was made of the *attar* of roses by the mother of Nurjahan and another perfume called *chua* were much in demand.[92] The ladies of the harem had great love for flowers and scents. Manucci describes the expanses of the harem on scents, flowers, rose water, scented oil distilled from different flowers as 'extraordinary', he writes,

The expenses of the mahal are extraordinary, for they never amount to less than a carol (karor) of rupees – that is 10 millions of rupees...The above expenditure will not appear incredible, when we consider that all persons in India being extremely choice about, and fond of scents and flowers, they disburse a great deal for essences of many kinds, for rose water and other scented oils distilled from different flowers.[93]

The use of perfumes also had a religious sanctity for Muslims therefore, as long as this segment of population enjoyed affluence or even purchasing power, perfume would not with standing its essential nature, continued to command a certain domestic market.[94]

Intoxicating liquors were consumed by a large section of the population. The fact is that, Indians both Hindus as well as

Muslims were addicted to the use of intoxicants in spite of the rigid requirement of their religion that they should not touch it.[95] The common and perhaps the cheapest drink was the *tari* or juice of palm, date or coconut.[96] Manucci writes that it was taken by Indians in place of wine.[97] Pleasant in taste and flavour, it was taken with pleasure throughout India. Cocoa juice was the principal ingredient for the preparation of a liquor which *"drinks as deliciously as wine"*. Indians particularly the Goanese liked it much and drank it like water.[97] Some superior kinds of wines were imported from the foreign countries like Portugal and Persia. Persian wines made from grapes were the favourite drink of Mughals.[99]

Indeed in the course of time, the demand for these preserved items increased so much that the state found it in its own favour to promote the industry of fruits and flowers preservation. Their manufacture received a new stimulus from the Mughal emperors. The history of perfumery industry in India can be traced from Harsha's (606-647 AD) period onwards. In Turkish period, royalty and nobility took interest in its manufacture. But it was the Mughal period when perfume industry received considerable royal patronage. Mughal emperors like Humayun and Akbar promoted this industry.[100] Akbar created a special department for perfumery called *'khushbu khana'* under the charge of Shaikh Mansur. Akbar encouraged this department for religious motive.[101] Father Monserrate has also made a mention of the *khushbu khana* of Akbar. He writes:

The common report of king's (Akbar) extraordinary kindness towards the priests... When they told king of this plan and its purpose, he immediately ordered the ointments, perfumes and very numerous jars of scented water to be conveyed out of that house into another; for the house (which the priest wanted) was used for manufacturing or storing of ointment and perfumes.[102]

Abul Fazl gives a list of 34 varieties of perfumes along with their prices which were available in the market. Those were:[103]

	PERFUMES	PRICES
1.	**Amber-i-ashhab**	1 to 3 muhurs, per tola
2.	Zabad (civet)	½ R. to 1 M. per tola
3.	Musk	1 to 4½ R., per tola
4.	Lignum aloes Hind. Agar	2 r to 1 m. per ser.
5.	Chuwa (distilled wood of aloes)	1/8 r. to 1 r., per tola
6.	Gaura	3 to 5 r. per tola
7.	Bhimsini camphor	3 r. to 3 per tola
8.	Mid	1 to 3 r. per tola
9.	Zafaran	12 to 22 r. per ser
10.	Zafaran-i-kamandi	1 to 3m. do
11.	Zafaran (from Kashmir)	8 to 12 r., do
12	Sandal wood	32 to 55 r., per man
13	Nafa-yi mushk	3 to 12 m., per ser
14	Kalanbak (calembic)	10 to 40 r per man
15	Silaras	3 to 5 r., per ser
16	Amber-i-ladan	1½ to 4 r., do
17	Kafur-i-china	1 to 2 r. do
18	Araq-i-fitna	1 to 3 r. per bottle
19	Araq-i-bed-i-mushk	1 to 4 r., do
20	Rose water	½ to 1 r., do
21	Araq –i-bahar	1 to 5 r., do
22	Araq-i-chambeli	1/8 to ¼ r., do
23	Violet root	½ to 1 r., per ser
24	Azfar ut tib	1½ to 2 r., per ser
25	Barg-i-maj(brought from Gujrat)	½ to 1 r., do
26	Sugandh gugala	10 to 13 r., do
27	Luban	1/3 to 3 r per tola
28	Luban (other kinds)	1 to 2 r., per ser
29	Alak, Hind. Char	¼ to ½ r., do
30	Duwalak, Hind. Chharila	3 to 4 d. do

31 *Gehla*
32 *Sufd*
33 *Ikanki*
34 *Zurumbad*

Large tracts of lands were brought under the cultivation odoriferous plants around Champaner, Ahmadabad, Surat, and Sironj and perfume making industry developed in these towns and also at Nausari.[104] Agra was also famous for its scents especially china which was the rich perfume. It was a liquid of black colour.[105] Kamroop was also famous for the preparation of perfumes.[106] In Awadh, cultivation of sweet scented flowers like roses of various kinds such as *bela*, jasmine, etc. was under taken and rose scent, rose oil, rose water, *itr* were extracted from these flowers. Jaunpur was noted for perfume essence, fragrant oil, chiefly *bela* oil, *itr* of Qannauj was renowned throughout the country.[107] In addition to these Lahore, Balsar, Cambaya and Benaras were well known for rare perfumes.[108] Manucci has praised the jasmine oil of the Gwalior and regarded it as *"the best to be found in the kingdom"*.[109] Manrique has praised the rich scented oils of Mickapur in Bengal.[110] Although the indigenous perfumes were evidently not as yet comparable to the Persian perfumes, but the art which had been introduced during this period promoted its industry and a large number of artisans were employed. All over North India, the mosque and saint's tombs provided work for petty craftsmen who made rose water and other things.[111] In Jaunpur the artisans produced rose water, turbans, etc. for local elites.[112] A class of scent merchants, existed in Bengal and were known as *Gandha Baniks*.[113] Apart from perfumery; the industry for fruit preservation also gained impetus. Indeed in the course of time the volume of fruits produced in the country came to be so plentiful that at Ahmedabad and Thaneswar fruit conservation assumed the character of a regular industry carried on for the purpose of export.[114] At Thaneshwar conserves like *murabba*, *chutney* and *achar* of vegetables and fruits was a flourishing industry; mangoes, *hur* (myrobalans) were the chief articles of conserves.[115] Bernier writes that in Bengal people preserved citrons in large quantities.[116] Fruits were also preserved in form of pickles. Abul Fazl has given a list of pickles along with their prices which were available in market. Those were[117]:

	Pickles	Prices
1.	Sour limes, per ser	6 d.
2.	Lemon juice, do	5d
3	Sugarcane vinegar, do	1 d.
4	Mangoes in oil, do	2d.
5.	Do in vinegar, do	2 d.
6.	Lemons in oil, do.	2d.
7.	Do. In vinegar, do	2 d.
8.	Do. In salt, do.	1½ d.
9.	Do in lemon juice, do	3d
10.	Pickled bamboo, per ser	4 d
11.	Do. Apples, do	8 d
12	Do. Quinces, do	9 d
13	Do, Raisins and *munaqqa*, do	8 d
14	Do. Peaches do	1 d
15	Do. Karil buds (capparis). Do	½ d.

Also the production of intoxicating liquor provided employment to the people on large scale as Terry writes that the chief occupation of the Parsis was planting and husbanding vine and toddy trees.[118] Tavernier mentions that in Golkunda the women ran the *Tari* shops.[119] The fruit preservation gave an impetus to the promotion of new industries, which provided employment on a large scale and also helped to improve the economy of the empire. The fruits and flowers when preserved by the various techniques and methods. These processes not only kept fruits and flowers products in use for long duration but more importantly, they became a source of income and livelihood. Though the making of the pickles was most popular form of preservation, various other methods such as in forms of *murabbas* and marmalades also contributed to the preservation of the fruits. In the preservation process both the state and society were involved. Besides, flowers were used for making of the perfumes and scents *(itr)*, this process also widened the scope of flower culture.

REFERENCES AND NOTES

1. Sinha, B.P. *Readings in Kautilya's Arthsasthra*. New Delhi: Agam Publishers, 1976. p-5, Faruki, *Aurangzeb and His Times*, pp-481,482.
2. *Baburnama*, p-503; Ain Vol.1, p-72.
3. *Ain* Vol.1, p-67.
4. Bernier, p-438.
5. *Khulasat* in *Geographical Glimpses* Vol.3, p-72.
6. *Ain* Vol.1, pp-79-81.
7. *Ibid.*, p-81.
8. *Ibid.*, pp-86, 87.
9. *Khulasat* in *Geographical Glimpses* Vol.3, p-63.
10. *Baburnama*, p-503, f.n. 6; also see Elliot & Dowson, Vol.V, p-24.
11. *Ibid.*, p-503.
12. *Ain* Vol.1, p-72.
13. Manucci Vol. III, p-171.
14. The *Jahangirnama*, p-197.
15. Ansari. *European Travelers*, p-88.
16. Bernier, p-438.
17. *Baburnama*, p-511.
18. *Ibid.*, pp-511, 512.
19. Della Valle Vol. II, pp-233, 234.
20. *Ibid.*, p-234.
21. Zain Khan's *Tabqat-i-Baburi*, p-122.
22. Manucci Vol. III, p-177.
23. *Khulasat* in *Geographical Glimpses* Vol. III, p-72.
24. *Ibid.*, p-72.
25. *Ain* Vol. I, p-68.
26. *Tuzuk* Vol. I, p-423.
27. Bernier, p-249.
28. *Baburnama*, p-203; f.n. 7.
29. *Baburnama*, p-505.
30. *Ain* Vol.1, p-89.
31. *Ibid.*, p-89.
32. *Ain* Vol.1, p-67.
33. *Ibid.*, p-67.
34. *Ibid.*, p-67.
35. *Ibid.*, p-67.
36. *Ibid.*, p-92.
37. Manucci Vol. III, p-171.
38. *Ibid.* pp-78, 79.

39. *Ibid.* p-79.
40. *Ibid.* p-79.
41. *Ibid.* p-79.
42. *Ibid.* p-79.
43. *Ibid.* p-79.
44. *Ibid.* p-80.
45. *Ibid.* p-80.
46. Meera Nanda. *European Travel Accounts*, p-105.
47. *Baburnama*, p-505.
48. *Tuzuk* Vol.1, p-6.
49. Manucci Vol. III, p-176.
50. *Ibid.* Vol. I, p-68.
51. Pelsaert, p-65.
52. *Ain* Vol. I, pp-79-81.
53. *Tuzuk* Vol.1, pp-270, 271; Also see The *Jahangirnama*, p-163.
54. Manucci Vol.1, pp-158, 159.
55. Chandra Pant. *Nurjahan and her family*, p-119.
56. *Baburnama*, p-514.
57. *Ain* Vol. I, pp-83-87.
58. *Ibid.* p-85.
59. *Ibid.* p-86.
60. *Ibid.* p-87.
61. *Ibid.* p-87.
62. Tavernier, p-15.
63. *Baburnama*, p-505.
64. *Ain* Vol. I, p-75.
64a. *Ibid.* p-74
65. *Baburnama*, p-509.
66. Manucci Vol. III, pp-176, 177.
67. *Ain* Vol. I, p-75.
68. *Ibid.* p-75.
69. *Baburnama*, p-509.
70. Tavernier Vol. I, p-194.
71. De laet, p-24.
72. *Aurangzeb in Kashmir*; p-78.
73. Ansari. *European Travellers*, p-94.
74. *Ibid.* p-100.
75. Careri. *A Voyage*, p-200.
76. Bernier, p-253; f.n. 3.
77. See *Ain* Vol. I, p-67; Also see Manucci Vol. III, p-171.
78. Pran Nath Chopra. *Some aspects of Society and Culture*, pp-40, 41.

79. *British Social Life in India*, p-13.
80. P.N.Ojha. *North Indian Social Life during the Mughal Period*, p-7.
81. P.N.Chopra. *Some Aspects of the Social Life*, p-37.
82. *Ibid.* pp-17, 18.
83. *Ibid.* p-18.
84. *Ibid.* p-18.
85. Pelsaert. p-65.
86. *Ain* Vol. I, p-79.
87. P.N.Chopra. *Some Aspects of the Social Life*, p-18.
88. P.N.Ojha. *op.cit.*, p-32.
89. *Ibid.* p-32.
90. P.N.Chopra, *op.cit.*, p-16.
91. *Ain* Vol.1, p-78-79.
92. Srivastava, M.P. *Policies of the Great Mughals*. Allahabad: Chugh Publications, 1978 p-47.
93. Manucci Vol. II, pp-315, 316.
94. Naqvi. *Urbanization and Urban Centres*, p-44.
95. D.Pant. *The Commercial Policy of the Moguls*, pp-19, 20.
96. *Baburnama*, p-509, Ain Vol.1, p-75.
97. Manucci Vol. III, p-177.
98. P.N.Chopra. *op.cit.*, p-48.
99. D.Pant. *op.cit.*, p-20.
100. M.P.Srivastava. *Policies of the Great Mughals*, p-46.
101. *Ain* Vol. I, p-78.
102. Father Monserrate, p-58.
103. *Ain* Vol. I, pp-80-81.
104. Naqvi. *Urbanization and Urban Centres*, p-44-45.
105. M.P.Srivastava. *op.cit.*, p-47.
106. Gulam Hussain Zaidpuri's *Riyaz-us-Salatin in Geographical Glimpses*, Vol. III, pp 53, 54.
107. *Journal of Pakistan Historical Society*, p-309.
108. P.N.Chopra, *op.cit.*, p-17.
109. Manucci Vol. I, p-68.
110. Meera Nanda. *European Travel Accounts*, p-105.
111. C.A.Bayley, p-129.
112. *Ibid.* p-133.
113. K.M.Ashraf. *Life and Conditions*, p-121.
114. Naqvi. *Urbanization and Urban Centres*, p-43.
115. Naqvi. *Upper India [1556 – 1803]*, pp-52-53.

116. Bernier, p-438.
117. *Ain* Vol.1, p-67.
118. Ansari. *European Travelers*, p-98, 99.
119. Tavernier, p-128.

APPENDIX - I

Glossary

1	*Abjosh*	-	Large raisins
2	*Abpashi*	-	A festival during which rose water was squirted on participants.
3	*Afonso*	-	The famous grafted mangoes created by Portuguese.
4	*Alu-balu*	-	A kind of plum.
5	*Ambah*	-	Mango
6	*Ambli*	-	Tamarind
7	*Araq*	-	Spirit
8	*Arra-kash*	-	Who used to saw beams.
9	*Bagh*	-	Garden
10	*Baolis*	-	Stepped wells
11	*Baradari*	-	A canopied building with 12 open doors on sides. It is a typical Hindu structure.
12	*Ber*	-	Lote fruit
13	*Briksha*	-	Tree
14	*Buyutat/karkhanas*	-	Workshops
15	*Chambeli*	-	Jasmine
16	*Champaka*	-	Michelia champca
17	*Charbagh*	-	Four fold plot, a garden with rising terraces.
18	*Chaukhandi*	-	An open four side pavilion, either a free standing structure or a pavilion mounted on an elephant.
19	*Chhatris*	-	Canopies

20	*Coquinhos*	-	Little Coconuts
21	*Daftar yu Sunluq*	-	Account rolls
22	*Jaman*	-	Eugenia Jambolana
23	*Jashans*	-	Social parties
24	*Kadamba*	-	Anthocephalus Cadamba
25	*Kadli*	-	Banana
26	*Kadhil*	-	Jackfruit
27	*Kasa-I-ghichaks*	-	A kind of violin
28	*Kanwal*	-	Lotus
29	*Khas*	-	A kind of grass which emits fragrance, especially when it is wet in water.
30	*Khilat-i-khasa*	-	A special robe given to the nobles by Emperors.
31	*Khurma*	-	Date palm
32	*Kishmish*	-	Raisins
33	*Khayaban*	-	Avenues
34	*Kharbuza*	-	Melons
35	*Khubani*	-	Dried apricots
36	*Lakhira*	-	Who varnished the reeds
37	*Mewah khana*	-	Special department for fruits
38	*Mewa-i-Shirin Hindi*	-	Sweet fruits
39	*Mewa-i-khushki*	-	Dried fruits
40	*Mewa-i-tars*	-	Sour fruits
41	*Mewa-i-Turan*	-	Central Asian Fruits
42	*Mighush*	-	Fruits some what with acid
43	*Naghzak*	-	Mango
44	*Nargil*	-	Coconut
45	*Nishkar*	-	Sugarcane
46	*Pan*	-	Betel leaf
47	*Pan-chakki*	-	Wind mill
48	*Patoles*	-	A kind of silk cloth, very soft

49	*Paywandi*	-	Grafting
50	*Payalas*	-	Cup
51	*Piyadah*	-	Foot soldiers
52	*Qand*	-	Sugar
53	*Qara-qashuq*	-	Black spoons made of coconut shells
54	*Rayhan*	-	Sweet basil
55	*Singhara*	-	Water chest nut
56	*Shahalu*	-	Cherry, name given by Akbar to the Cherry.
57	*Shah-toot*	-	Mulberry
58	*Sharifa*	-	Custard-apple
59	*Safaida*	-	A light yellow sucking variety of mangoes
60	*Serias*	-	Inns
61	*Tarbuz*	-	Water melon
62	*Udyan*	-	Garden
63	*Zafran/Kesar*	-	Saffron

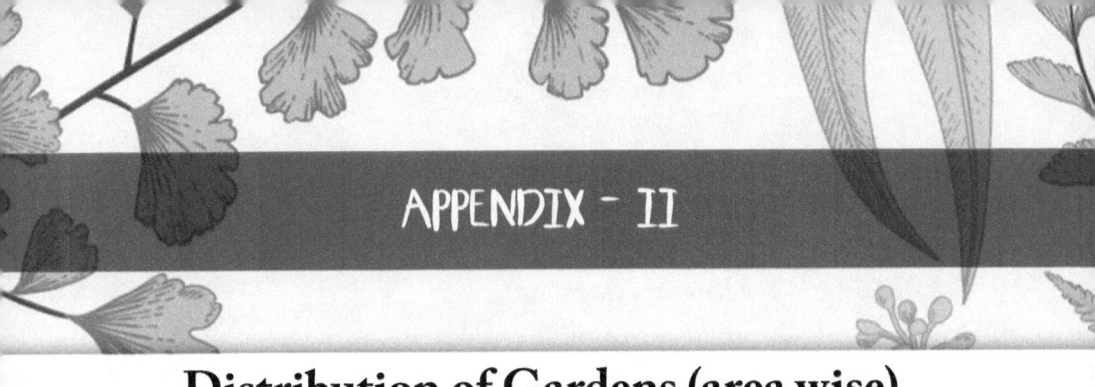

APPENDIX - II

Distribution of Gardens (area wise)

Area	Gardens
Kabul	- *Bagh-i-Wafa (Babur's tomb)*
Kashmir	- Nasim Bagh
	Shalamar Bagh
	Nishat Bagh
	Chasma Shahi
	Pari Mahal
	Achabal
	Vernag
Rajauri	- *Wah bagh*
Lahore	- *Shalamar Bagh*
	Shahdara
	Fort Gardens
Kalka	- *Pinjaur*
Delhi	- *The Red Fort*
	Humayun;s tomb
	Roshanara's Garden (Pavilion only)
Agra	- *The Taj Mahal*
	Anguri bagh in the Fort
	Aram Bagh
	The Tomb of I'timad-ud-Daula
	Akbar's tomb, Sikandara
	The Sultana's garden at Fatehpur Sikri
Bharatpur	- *Deeg (in the Mughal tradition)*
Jaipur	- *Amber (two gardens, Mughal-Rajput)*
Ajmer	- *Shahjahan's pavilions.*

Udaipur	-	*Sahelion-ki-bari (Mughal by derivation)*
		Lake palace landscape
		(Rajput – Mughal)
Allahabad	-	*Khusrau Bagh*
Mandu	-	*Nilkanth pavilion*
Aurangabad	-	*Mausoleum of Rabia Darauni*
		Pan-chakki mill

APPENDIX - III

Trees, Plants, Flowers, Fruits – their significance

	Trees and Plants	- Significance
1.	*Sygwan or teak wood* *Mahuwa wood* *Talip-pat* *Sheesham wood* *Siso-wood* *Pine-wood* *Nazhu(Jidh) wood* *Dasang (kari) wood* *Ber wood* *Mughilan wood* *(Babul)* *Sirs* *Dayar* *Bakayin* *Mango wood*	All these kinds of woods were used for the construction of buildings, bridges, means of transport such as carts, boats, ships, planks etc. Wood was also used for making articles such as furniture – chairs, tables, tepoys, book cases, bed steads, boxes, spoons, etc.
2.	*Egg plum or beir tree* *(ziziphus jujuba)*	- Bark was utilized in the preparation of medicine for frightful severe disease – the tape worm, also used for making strong and durable silk – *tusser*.
3.	*Jamrool tree* *(eugenia alla or aquea)*	- Bark was thought a sovereign remedy for apthae in children.

4.	*Palm tree*	- Branches were used for making mats, leaves for making fans, mats, sun shades, small baskets and other curiosities.
5.	*San (hemp)*	- Bark was used for making strong ropes
6.	*Tesi tree*	- Branches were used for making mats.
7.	*Guava tree*	- Wood was used for making gun-stock.
8.	*Palmyra-palm(Tar)*	- Leaves were utilized for writing, also a liquor called Tari was extracted from this tree, fruit had medicinal value – eating it was useful in clearing the sight.
9.	*Tuz tree*	- Bark was used for writing purposes *(Bhoj-pattra)*
10.	*Date palm (Khurma)*	- Yielded intoxicant liquor
11.	*Mulberry tree*	- Facilitated the emergence and expansion of sericulture in India.
12.	*Babul (wood)*	- Ashes were used for refining adulterated silver.
13.	*Tar tree*	- Leaves were used for writing purposes.

	Fruits	**Significance**
1.	*Amla*	- Has the medicinal properties
2.	*Ananas (pine-apple)*	- Distilled juice of ananas was useful in dissolving stones in kidneys and bladder.
3.	*Kaju (cashewnut)*	- The juice of cashewnut was used to clear all obstruction in the breast.

4.	*Coconut*	- Coconut was used for the preparation of antidote against poison, water of coconut was used for inflammation of liver, the kidneys and bladder and helped to increase urination. Shells of coconut were used for making dishes, vessels, spoons, cups, buckets, hookah, bowls, lamps, etc. the ropes of the ships and boats were made from the husks of coconut, oil was extracted from nut of coconut, useful for burns and ulcers.
5	*Lemon*	Fiber of lime in boiled water was a cure for poisoning. Lemonade was used to cure prickly heat during summer.
6.	*Mango*	- Old kernels were used as medicine, and mango was also used as mordanting agents in textile industry.
7.	*Mirobalanos quebulos (Chebulic myrobalan)*	- Medicinal values.
8.	*Guava*	- Medicinal value
9.	*Plantain (Banana)*	- Dressings for blisters, or as a covering for shaven head in case of brain fever.
10.	*Pumple-nose (citrus decumanus)*	- Sherbet was most grateful drink to sick.

11.	*Pomegranate*	-	Red colour from fruit and flowers used as dyeing agent in textile industry.

	Flowers		**Significance**
1.	Such as *Bholsari, Sewti, Chambeli, Champa, Nargis, mongra, Kewara, Jasmine, roses, gul-i-heena,* and so on.	-	For making perfumes and oils
2.	*Egyptian willow flower (bed-i-mushk)*	-	Distilled water useful for all kinds of fever caused by heat.
3.	*Rose*	-	rose water used for making cough medicine
4.	*Henna, Saffron*	-	As dyeing agents in Textile Industry
5.	*Ratan-mala flower*	-	Juice furnished a fast dye for stuffs.
6.	*Champala*	-	Provided red dye

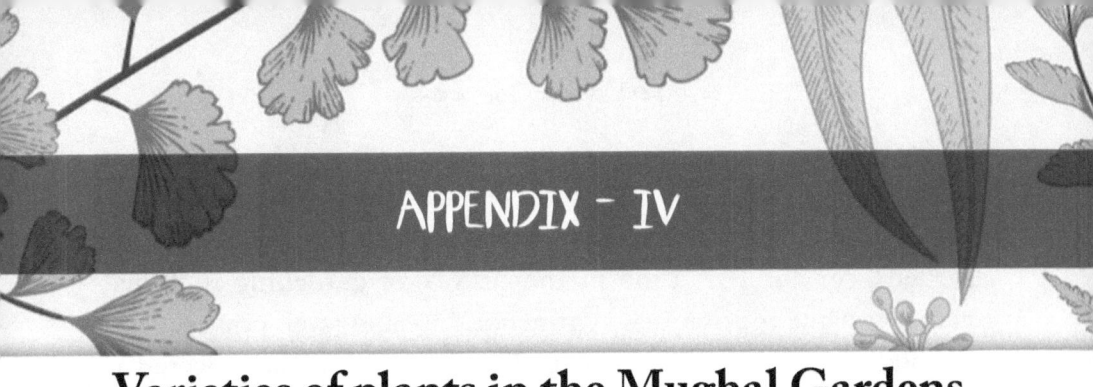

APPENDIX - IV

Varieties of plants in the Mughal Gardens

From the contemporary Mughal sources it is evident that their gardens were embellished with choicest trees, flowers and fruits. The memoirs of Babur and Jahangir are filled with description of fruits and flower trees in their gardens. The planting can also be guessed at from the contemporary paintings. In the paintings of the period trees and flowers abound especially the *Chinar,* the cypresses, spring flowers scatter the gardens and branches of fruit blossom overhang the walls. Some information is also available from the accounts of the European travellers in the Mughal empire. The planting of the trees was held in great regard by the Mughals and in their gardens they planted plants and flowers of every kind. While describing about his garden of *bagh-i-wafa,* Babur mentions the fruit and flowers which were grown in this garden:

I laid out the four-gardens, known as bagh-i-wafa (garden of fidelity)… there oranges, citron, pomegranates grow in abundance…I had plantains brought and planted there; this did very well. The before I had the sugarcane planted there; it also did well; some of it was sent to Bukhara and Badakshan.

Names of the numerous trees have been mentioned in the memoirs of Babur and Jahangir. Ain-i-*Akbari* of Abul Fazl gives a long list of the fruit and ornamental trees grown in the Mughal gardens. A builder rather than a gardener Akbar had nevertheless, some hand in the making of gardens and had trees and flowers of all kinds imported and planted at Agra and Fatehpur Sikri. Peter Mundy writes that in the gardens of Agra many fruit trees such as apple trees, mango trees, orange trees, mulberry trees, coconut trees, fig trees, plantain trees and cypresses were planted. Jahangir with nature as his passion, carried the process step further. In his gardens, for the first time in the history of gardening in India, nature became the dominant influence. Site and level, prospect and orientation, plantation and above all water in all its forms were the real constituents in the gardens. During Jahangir's reign numerous fruits such as melons, mangoes and many others grew well in Agra and its neighbourhood. Cultivation of *ananas* is also reported from the royal garden at Agra. Jahangir writes in his memoirs:

Among the fruits one which they call ananas (pineapple), which is grown in Frank-ports, is of excessive fragrance and fine flavour. many thousands are produced every year now in the gulafshan garden at Agra.

In the gardens of Kashmir trees such as oriental plane, cypress, willow and poplar were widely planted along with numerous fruit trees. Bernier who visited the garden of Achhabal in 1665 reports:

The garden itself is very fine, there being curious walks in it, and a store of fruit bearing trees, of apples, pears, prunes, apricots, and cherries, and many jets of water of various figures…

In present day Kashmir, the abundance of fruits of all kinds, and particularly the excellence of walnuts tallies very closely with the description of Mughals. Sweet cherry was not grown in kashmir prior to Akbar's reign, but it was introduced during his reign in Kashmir. An important feature of every Mughal garden in Kashmir and of the landscape of the Dal lake was the heavy solidity of the *chinar* trees (plantanus orientalis). Not only did they give weight, shape and form within the gardens, but their their scale was a vital element in relating the gardens to the vast scale of lake and the mountains. In Kashmir, Shahjahan planted four *chinar* trees on an island on Dal lake and that island came to be known as *char-chinar*. At the site of Naseem *bagh*, he also got planted hundreds of *chinar* trees on a regular grid. The sandal trees also flourished in the Mughal gardens. The emperors were very particular in the import of fruit trees, while they also followed the age old tradition of planting the shade giving trees. The planted avenues along the roadsides. At halting places the trees were often extended four square, so as to provide shade from all points of the compass. Peter Mundy tells us that when Jahangir planted the road side avenues stretching out from Agra, he used *neem* (margosa), *pipal* (ficus religious), *dhak* (butea frondosa), and *bahar* (ficus Indica) for the use of the travellers and for the shade in hot weather. shahjahan is said to have imported fruit trees from Kabul and Kandhar and got them planted at the shalamar garden, Lahore. At this garden there were continuous flower beds with plane trees and aspens at intervals. Shahjahan is recorded as having himself planted an aspen between two planes. It was Lahore that William Finch of East India Company gave one of the most detailed descriptions of actual flower to fruit planting:

...adjoining to this is a garden of king in which are very good apples but small, toot (tut-mulberry), white and red, almonds, peaches, figs, grapes, quinces, oranges, lemons, pomegranates, stock yellow flowers (the white stock, mathiola incana), marigolds, wall-flowers, ireos (the florentine iris), pinkles white and red, with diverse sorts of Indian flowers.

Red fort at Delhi had two beautiful gardens, the *Hayat Bakhsh* or life giving garden and *Mahtab Bagh* or moonlight garden. Richly planted, the colours of *Hayat Bakhsh* were predominantly crimson and purple and the scent prevailed the garden. The *Mahtab bagh*, by contrast, was planted with pale yellow flowers only like jasmine, tuberoses, lilies and narcissus, while both the gardens were enclosed by rows of cypresses. In fact the Mughals understood and used their native flora, creating gardens which were the reflection of both their way of life and condition of their country. The Mughals mostly planted spring-flowering trees, shrubs and herbs in their gardens. They grew white, purple and mauve iris near lilac bushes, daffodils and narcissi under apple and quince trees, and tulips under pear and plum trees in their Kashmir gardens. In summers they grew roses, carnations, jasmines, hollyhocks, peonies and delphiniums. Some times the flowers of one variety were massed in a garden thus creating a beautiful colour effect.

APPENDIX - V

Rajput Gardens

The history of gardens closely follows the history of arts like painting and architecture. With the decline in patronage at Delhi, the artists trained in the Mughal style of painting migrated to the states of Rajasthan, central India and western Himalayas-where they evolved new styles. There they were welcomed and employed by the Rajputs who inherited apart of Mughal-Persian culture together with Persian court language. Similarly the Mughal garden started a new lease of life in Rajasthan and were transformed into Rajput gardens. The Rajput states like Amber (Jaipur), Bikaner and Jodhpur were in intimate contact with the Mughal court and absorbed its culture. The eclecticism is evident in the architecture of their palaces, the style of their paintings, clothes of their women and the style of their gardens. All were influenced by the Mughal tradition. They were wise enough to appreciate the virtues of the formal garden of irrigation in combating the heat of India by the use of wide canals. Therefore the Rajput gardens represented a cultural synthesis and a mutation.

But on the other hand conservative Mewar was insular and rejected Muslim influences. This is reflected in its art and architecture including gardens which are more indigenous in character. In gardens, the Rajput heritage is perhaps most clearly seen at Amber near Jaipur. The gardens with the romantic lake and hill side settings, their strange combination of the exact and the picturesque perfectly express the union of widely repeated ideals. The palace of Amber was originally built by Raja Man Singh (A.D. 1590-1615), who was the foremost general of Akbar. But it was completed by Jai Singh II (A.D. 1699-1743) in the early eighteenth century. Below the palace is a large lake with a beautiful garden. The garden is in three terraces with parterres of different shapes. Looked at from the windows of the palace it appears as mosaic with many geometrical patterns. Planted with flowers of various colours, the garden looked like a colourful carpet. There is another garden in the front of ladies palace which was cooled by the spray of the fountains. When the garden was in its full glory it was a delightful place scented with the fragrance of jasmine, *champa* and *moghra*. The stone parterres in the two important Amber gardens are based upon stars which were held in high esteem by Seljuk Turks, for whom they stood for life and for man's intellectual powers. It may be significant to note that the immigrants of Turkish origin had once taken refuge in Rajputana.

Jodhpur was also famous for its gardens. Raja Abhai Singh (A.D. 1724-1749) was builder of palaces and gardens. He laid gardens at Jodhpur and Mandor. Mandor garden was visited by Tod in A.D. 1828. He narrates:

I now retired to the palace and gardens built by Raja Ajeet…the gardens though not extensive, as may be supposed, being confined within the adamantine walls reared by the hands of nature, must be delightfully cool even in summer. Fountains, reservoirs and water courses are every where interspersed …Some attention was paid to its culture, besides many indigenous shrubs, it boasted of some exotics. There golden champa, the pomegranate, at once rich in flower and fruit, the apple of Sita or Sitaphal…a delicious species of the plantain…the moghra, the chameli or jasmine…

Another notable garden in Rajasthan is at Deeg in Bharatpur state. The Raja of Bharatpur state laid out a garden at Deeg in 1725, embodying the paradise tradition as adopted to the hot land of Hind. So well attuned to the Indian climate and way of life in this tradition that one may hope to see it undergo a new renaissance. Dee with its fountains, water courses, parterres and noble buildings which were richly ornamented with precious stones rivalled the Mughal gardens of Agra. At Udaipur two island palaces upon lake Pichola, namely

Jagniwas and *Jagmandir*, seems to float like pearls on the smooth water. These were built by Rana Jagat Singh (A.D. 1628-1652), a patron of paintings and builder of palaces and gardens. Tod writes that in these palaces:

Parterres of flowers, orange and lemon groves, intervene to dispel the monotony of the buildings, shaded by the wide spreading tamarind and magnificent ever green khirnes. While the graceful palmyra and cocoa wave their plum like branches over the dark cypress or cooling plantain

Bundi with its lakes, gardens, forests and palaces was another cultural centre in eastern Rajasthan. One of the Bundi ruler, Rao Chattar Sal (A.D. 1652-1658) was made the governor of the Imperial capital of the Mughals that is of Delhi by Shahjahan. On account of Rao Chattar Sal's contact with the Mughal court, the arts of painting and gardening reached Bundi.

As the Mughal power waned, the Punjab hill-states in the western Himalayas became the scene of the cultural renaissance besides Rajasthan. The Hindu artists trained in the Mughal style reached a tiny state of Guler, now in Kangra district of Himachal Pradesh. Under the patronage of the art loving ruler, Raja Govardhan Chand, they developed the Kangra style of painting. In Kangra paintings we perceive deep love of nature. The Kangra style painting also reached Chamba and flourished under the patronage of Raj Singh (A.D. 1755-1794). along with the art of painting came the art of gardening. A Mughal style *charbagh* with two pavilions was laid down at Raj nagar. Though the garden has been disappeared now, but we find it recorded in a charming painting, in the collection Louvre, Paris. Another patron of paintings and gardening was Raja Sansar Chand (A.D. 1765-1823) of Tira Sujanpur. At Alampur he planted a garden, which is said rivalled the shalamar garden of Lahore. the ruins of this garden structure can still be seen scattered in the wheat fields. The Rajput gardens of the Himalayan states were designed for the pleasure of the rulers. Numerous paintings of the period show the couples in the garden pavilions. By the middle of the nineteenth century the art of the miniature paintings languished from Kangra valley. The *charbagh* type of garden also disappeared along with the art of painting.

BIBLIOGRAPHY

Primary Sources

Afif, Shams Siraj.	*Medieval India in Transition - Tarikh-i-Firozshahi.* A first hand account, R.C.Jauhari. New Delhi: Sundeep Prakashan, 2001.
Ahmad, Khwaja Nizamuddin.	*The Tabqat-i-Akbari* Vol. I, Tr. and annotated by Brajendranath and revised and ed. by Baini Prasad, Jammu: Jay Kay Book House, 1994.
Alberuni,	*Alberuni's India (Tarikh-i-Hind)* Trans.by Dr. Edward C.Sachau. New Delhi: Rupa & Co., 2002.
Allami, Abul Fazl	The *Ain-i-Akbari*, Vol. II & III Eng.Trans. by Col.H.S.Jarret, revised and further Annotated by Sir J.N.Sarkar, reprint Delhi: Low Price Publications, 1994.
Allami, Abul Fazl.	The *Ain-i-Akbari*, Vol. I, Trans. by H.Blochmann, reprint Delhi:Low Price Publications, 1997
Allami, Abul Fazl.	The *Akbar Nama*, Vols. I, II & III, Eng.Trans. by H.Beveridge. reprint Delhi: Low Price Publications, 2017.
Anonymous,	*Baharistan-i-Shahi*, A Chronicle of Medieval Kashmir. trans. by K.N.Pandit Calcutta: Firma kilm Private Limited,1991.
Ansari, M.A.	*Geographical Glimpses of Medieval India* Vols. I, II and III, Intr. With Eng.abstracts by Jaweed Ashraf, Tasneem Ahmad, Delhi: Idarah-i Adabiyat-i Delli, 1989.

Ansari, M.A.	*Administrative Documents of Mughal India.* Delhi: B.R. Publishing Corporation,1984.
Ansari, M.A.	*European Travellers under the Mughals.* Delhi: Idarah-i Adabiyat-i Delli, 1975.
Babur, Zahiruddin Mohammad.	*Baburnama,* vols. I & II in one format, eng. trans. by A. Beveridge. reprint Delhi: Low Price Publications, 1997
Badauni, Abdul Qadir.	*Muntakab-ut-Tawarikh* Vol. I, trans. G.S.A.Ranking, Patna1973, Vol. II, Trans.W.H.Lowe, Patna, 1973, Vol. III, Trans. Sir Walseley Haig, Patna: Academica Asiatica, 1973.
Battuta, Ibn.	*The Travels of Ibn Battuta* Vol. III A.D. 1325 – 1354. Trans. with versions and notes from the Arabic text ed. by C.Defremery and B.Sanguinetti by H.A.R.Gibb, London: Hakluyt Society, 1971.
Batuta, Ibn.	*Travels in Asia and Africa 1325-1354.* Edited by Sir E.Denison Ross and Eileen Power, trans. H.A.R.Gibb. Reprint Delhi: Pilgrim Books, 1988.
Beal, Samuel.	*SI-YU-KI:Buddhists records of the western world.* Translated from the Chinese of Hieun Tsiang, AD 629. reprint New Delhi: Munshiram Manohar Lal Publishers, 2014.
Bendrey, V.S.	*A Study of Muslim inscriptions* with special reference to the inscriptions published in the Epigraphia Indo- Moslemica (1907-1938). Together with the summaries of inscriptions chronologically arranged. Forwarded by Khan Bhadur Prof. Shaikh Abdul Kadir Sarfaraz, Delhi: Anmol Publications, 1985.
Bernier, Francois.	*Travels in the Mogul Empire.* Eng.Trans. by A.Constable, Delhi: S.Chand & co., 1968.

Bernier, Francois. *Aurangzeb in Kashmir (Travels in the Mogul Empire)* – Eng. Trans. by Irving Brook, ed. by D.C.Sharma, Delhi: Rima Publishing House, 1988.

Bhandari, Sujan Rai. *Khulasat-ut-Tawarikh*, ed. Zafar Hasan, Eng. trans. by J.N.Sarkar, Delhi: J&sons 1918.

Careri, J.F.Gemelli. *A Voyage round the world by John Francis Gemelli, Careri*, ed. by Surendranath Sen, Indian travels of Thevenot and Careri, 1st edition New Delhi: National Archives of India, 1949.

Carre, Abbe. *The Travels of the Abbe Carre in India and the Near East – 1672 to 1674* in 3 Vols. Translated from the manuscript journal of his travels in India by Lady Fawcett, ed. by Sir Charles Fawcett, Sir Richard Burn. New Delhi: Asian Educational Service, 1990.

De Laet, *The Empire of the Great Mogol (De Imperio Magni Mogolis)*, DeLaet's description of India and Fragment of Indian History. Eng.Trans.by J.S.Hoyland, Annotated by S.N.Banerjee, New Delhi: Munshi Ram Manohar Lal Publishers, 1974.

Della Valle, Pietro. *The Travels of Pietro Della Valle in India* Vols.I & II ed. by Edward Grey from the old translation of 1664, by G.Havers, New Delhi: Asian Educational Services, 1991.

Dughlat, Mirza Muhammad Haidar. *The Tarikh-i-Rashidi*, Eng.Trans. by E Denison ross, ed. by N Elias, Delhi: Renaissance Publishing House, 1986.

Elliot and Dowson, *The History of India as told by its own Historians. Vol.V,* first edition Delhi: Kitab Mahal (WD.) Private Limited, 1964.

Father Monserrate, S.J.,	*The Commentary of Father Monserrate, S.J.* – on His Journey to the Court of Akbar, translated from Latin by J.S.Hoyland, Annotated by S.N.Banerjee, New Delhi: Asian Educational Service, 1992.
Finch, William.	*India as seen by William Finch.* ed.R.Nath (Historical Research documentation Programme, May 1995) Jaipur: 1995.
Foster, William (ed.)	*The Voyage of Thomas Best to the East Indies (1612-14),* NewDelhi: Munshiram Manohar Lal, 1997.
Foster, William (ed.)	*Early Travels in India (1583-1619).* London: Humphrey Milford, Oxford University Press, 1921
Foster, William (ed.)	*The Embassy of Sir Thomas Roe to India (1615-1619).* As narrated in his journal and correspondence. NewDelhi: Munshiram Manohar Lal, 1990.
Foster, William (ed.)	*The Journal of John Jourdan (1608-1617).* New Delhi: Asian Educational Services, 1992
Gul Badan begam,	*Humayun-nama (The History of Humayun)* trans. with introduction and notes by Annette S.Beveridge, 1st Indian edition. Delhi: Oriental Books reprint Corporation, 1983.
Jahangir,	*The Jahangirnama* – Memoirs of Jahangir, Emperor of India trans.,edited and annotated by Wheeler M.Thackston, New York: Oxford University Press, 1999.
Jahangir,	*Tuzuk-i-Jahangiri or Memoirs of Jahangir,* Eng. Trans. by Alexander Rogers, ed., by Henery Beveridge, Vol.I & II complied in one. reprint Delhi: Low Price Publications 1999.

Khan, Inayat.	*The Shahjahan-nama*, Eng.Trans. W.E. Begley and Z.A. Desai, New Delhi: Oxford University Press, 1990.
Khan, Khafi.	*Muntakhab-Al Lubab*, Trans. by Anees Jahan Sayed, Bombay: Somaiya Publications Pvt. Ltd., 1977.
Khan, Saqi Must 'ad.	*Maasir-i-Alamgiri*, A History of Emperor Aurangzeb-Alamgir (Reign1658– 1707A.D.) Eng. translation and annotated by Jadunath Sarkar (2nd Ed.), New Delhi: Munshiram Manohar Lal,1986.
Khan, Zain.	*Tabqat-i-Baburi*, translation and introduction by Sayed Hasan Askari, Annotation by B.P. Ambastha. Delhi: Idarah-i Adabiyat-i Delli, 1982.
Khwandamir,	*Qanun-i-Humayuni* or *Humayunnama*. Trans. with explanatory notes by Baini Prashad. Reprint, Calcutta: The Asiatic Society, 1996.
Manucci, Niccolao.	*Storia Do Mogor or Mogul India (1653 – 1708)* Vol.I, tr. by William Irvine, reprint Calcutta: Editions Indian,1965.
Manucci, Niccolao.	*Storia do Mogor or Mogul India* Vol.II, tr. by William Irvine, also included the 3rd part of the History of the Mogul by Nicolas Anouchy, Venetian, Calcutta: Editions Indian, 1967.
Manucci, Niccolao.	*Storia do Mogor or Mogul India* Vol.III, tr. by William Irvine, Calcutta: Editions indian, 1967.
Markham, Clements R.(ed.)	*The Voyages of sir James Lancaster to East Indies and Voyage of Captain John Knight (1606)* London: The Hakluyt Society,1877.

Martin, Francois. Memoirs	*Travels to Afirca, Persia and India.* Two Vols., ed. and Trans. from French by Aniruddha Ray, Calcutta: Subarnarekha,1990.
Mehta, Bal Mukund.	*Letters of a King-Maker of the 18th C. (Bal Mukund Nama).* Eng.trans.with introduction and notes by Satish Chandra, Bombay: Asia Publishing House, 1972.
Mukhlis, Anand Ram.	*Mirat-ul-Istilah,* Encyclopaedic Dictionary of Medieval India, tr. by Tasneem Ahmad, New Delhi: Sundeep Prakashan, 1993.
Mukhlis, Anand Ram.	*Safar-nama-i-Mukhlis in Geographical glimpses of Medieval India.* Vol.III by M.A.Ansari, Introduction with English abstracts by Jaweed Ashraf and Tasneem Ahmad, Delhi:Idarah-i Adabiyat-iDelli, 1989.
Nagar, Ishwar Das	*Futuhat-i-Alamgiri.* Trans. and edited by Tasneem Ahmad, Delhi: Idarah-i Adabiyat -i Delli, 1978.
Palsaert, Francisco.	*Jahangir's India, the Remonstrantee of Francisco Pelsaert* - trans. from Dutch by W.H.Moreland and P.Geyl, Delhi: Idarah-i Adabiyat-i Delli, 1972.
Qandhari, Muhammad Arif.	*Tarikh-i-Akbari,* An annotated trans. with Intro. by Tasneem Ahmad, forward by Irfan Habib, Delhi: Pragati Publications, 1993.
Razi, Amin Ahmad.	*Haft Iqlim in Geographical Glimpses of Medieval India* Vol.II by M.A.Ansari. First edition Delhi: Idarah-i Adabiyat-i Delli,1989.
Razi, Aqil Khan.	*Waqiat-i-Alamgiri.* ed.by Khan Bahadur Maulvi Haj Zafar Hasan, Delhi: Mercantile, Print, Press, 1946.

Tavernier, Jean Baptiste.	*Travels in India* Vol.I, II & III, Eng.Trans. by V.Ball, ed. by William Crooke, Delhi: Oriental Books reprint Corporation. 1977.
Terry, Edward.	*A Voyage to East India: reprinted from the edition of 1655. With copper plates.* London:Gale Ecco Print Editions,1777
Tirmzi, S.A.I.	*Mughal Documents(1526–1627).* New Delhi: Manohar Publishers,1989.
Tod, James	*Annals and Antiquities of Rajasthan.* Vols. I&II. New Delhi: Rupa Publications India Private Limited,2011
Verma, B.D.	*Akhbarat:News letters of the Mugal Court,* Reign of Ahmad Shah, 1751-52. Bombay:1949
Yadgar, Ahmad.	*Tarikh-i-Salatin-i-Afghana in The History of India as told – By its own Historians* by Elliot and Dowson Vol.V. First edition Delhi: Kitab Mahal (WD.) Private limited,1964.

MODERN WORKS

Alam, Muzaffar.	*The Crisis of Empire in Mughal North India Awadh and Punjab (1707 – 1748).* New Delhi: Oxford University Press, 1986.
Alavi, Rafi Ahmad,	*Studies in the History of Medieval Deccan.* Delhi: Idarah-i Adabiyat-i Delli, 1977.
Anonymous	*"Military Defence of Our Empire in the East,"* Calcutta Review Vol.II, Oct. Cambridge University Press, Dec.1844.
Ansari, M.A.	*Social Life of Mughal emperors (1526 – 1707).* New Delhi: Geetanjali Publishing House, 1974.
Ashraf, K.M.	*Life and conditions of the People of Hindustan.* 3rd edition New Delhi: Munshiram Manohar Lal, 1988.

Augustus and Beveridge,	*The Emperor Akbar.* Patna: Associated Book Agency, 1973.
Bamzai, P.N.K.	*A History of Kashmir: Political, Social, Cultural, from the earliest times to present day.* Delhi: Metropolitan Book co., 1962.
Basham, A.L.	*The Wonder that was India.* Calcutta: Rupa&co., 1996.
Bayly, C.A.	*Rulers, townsmen and Bazaars North Indian society,* In the age of British Expansion, 1770 – 1870. New Delhi: Oxford University Press, 1993.
Bhatnagar, G.D.	*Awadh Under Wajid ali shah.* Varanasi: Bhartiya Vidhya Bhawan, 1968.
Brooks, John.	*Gardens of Paradise (The History and Design of the Great Islamic Gardens).* London: George Weidenfeld & Nicholson Ltd, 1987.
Brown, Percy.	*Indian Architecture (Islamic Period).* reprint Bombay: D.B.Taraporevala sons & co. Pvt. Ltd., 1995.
Chandra, Satish	*Parties and Politics at the Mughal Court 1707–1740.* NewDelhi: People's Publishing House, 1972.
Chandra, Satish	*Medieval India from Sultanate to the Mughals* Part-I (Sultanate 1206-1526), Delhi: 1999.
Chopra, Pran Nath.	Some aspects of *Society and Culture during Mughal Age (1526–1707).*Agra: Shiv Lal Aggarwal & Co.Ltd., Educational Publishers, 1955.
Choudhary, Tapan Ray and Irfan Habib(ed.)	*Cambridge Economic History of India,* Vol.I, C.1200 – C.1750.reprint New Delhi: Orient Logman in association with Cambridge University Press, 1993.

Choudhry, S.C.Ray.	*History of Mughals* (Comprehensive History of India Vol.IV) (From 1526 – 1707 A.D) 2nd Ed., New Delhi: People's Publishing House, 1990.
Coomara Swamy, Ananda K., Sister Nivedita,	*Myths of Hindus and Buddhists.* republished Mumbai: Jaico Publishing House, 1999.
Crowe, Sylvia and Sheila Haywood	*The Gardens of Mughal India.* Delhi: Vikas Publishing House Pvt. Ltd., 1973.
Darbari, Neera	*Northern India Under Aurangzeb (Social and Economic conditions).* Meerut: Pragati Prakashan, 1982.
Dayal, Maheshwar.	*Rediscovering Delhi.* New Delhi: S. Chand& Co. Pvt.Ltd.,1975
Douie, Sir James.	*The Punjab, North-West Frontier Province and Kashmir.* Delhi: Low Price Publications, 1994.
Dr.Krusnachandra Jena	*Babur and Baburnama.* Delhi: 1978.
Early, Abraham.	*The Last Spring, Lives and times of the Great Mughals.* New Delhi: Viking(India), 1997.
Erskine.	*History of India Under Baber* Ist edition, Delhi: Atlantic Publishers & Distributors, 1989.
Farooque, Abul Khair Muhammad.	*Roads and Communications in Mughal India.* Delhi: Idarah-i Adabiyat-i Delli,1977.
Faruki, Zahiruddin.	*Aurangzeb and His Times.* Delhi: Idarah-i Adabiyat-i Delli,1972.
Findly, Ellison Banks.	*Nurjahan, Empress of Mughal India.* New Delhi: Oxford University Press,1993.
Fujiwara, Shinya.	*The Beautiful World,* Kashmir, Vol.60. 1st edition. Delhi: Allied Publishers,1982.
Habib, Irfan.	*Agrarian System of Mughal India (1556– 1707)* 2nd revised ed. New Delhi: Oxford University Press, 1999.

Habib, Irfan.	*An Atlas of the Mughal empire.* Political and Economic maps with detailed Notes, Bibliography and Index. New Delhi: Oxford University press, 1986.
Habib, Irfan.	*Essays in Indian History* towards Marxist perception. Reprint New Delhi: Oxford University Press, 2002.
Haig, Sir Wolseley.	*The Cambridge History of India,* Vol.IV, The Mughal Period, ed.by Sir Richard Burn, New Delhi: Cambridge university Press,1979.
Hasan, Mohibul.	*Babur.* NewDelhi: Manohar Publishers, 1985
Hassnain, F.M.	*The History of Jammu, Kashmir, Ladakh and Kishtwar.* Delhi: Rima Publishing House,1998.
Hutchinson, Lester.	*European Freebooters in Moghul India.* Bombay: Popular Press, 1964.
Irvine, William.	*Later Mughals* Vol.II 1719-1739 edited by J.N.Sarkar, Calcutta: M.C.Sarkar & sons, 1922.
Jaffar, S.M.	*Some cultural Aspects of Muslim Rule in India.* Delhi: Idarah-i Adabiyat-i Delli,1972.
Jauhari, R.C.	*Firozshah Tughlak* 1351–1388 A.D. Agra: ABS Publications,1968.
Kaul, Manohar.	*Kashmir, Buddhist and Muslim Architecture.* New Delhi: Sagar Publications, 1971.
Keene, H.G.	*Hand Book to Agra.* Calcutta: Thacker, Spink, 1888.
Khosla, Ram Prasad.	*Mughal Kingship and Nobility.* Delhi: Idarah-i Adabiyat-i Delli, 1976.
Koul, Pandit Anand.	*Archaeological remains in Kashmir.* Lahore: The Mercantile Press, 1935
Kulkarni, Ramchandra.	*Maharashtra in the Age of Shivaji.* New Delhi: Diamond Publications,2008

Kumar, Dr.Anil	*Asaf Khan and His Times*. Patna: K.P.Jayaswal research Institute, 1986.
Lawrence, Walter.	*The Valley of Kashmir*, Srinagar: Kesar Publishers, 1967.
Mackenzie, Gordon.	*A Manual of Kistna District in the Presidency of Madras*, compiled for the governed of Madras. Madras: Printed at Lawrence Asylum Press by W.H.Moore,1883.
Majumdar, R.C., J.N.Chaudhuri and S.Chaudhuri (ed.)	*History and Culture of Indian People*, Vol.VII, The Mughal Empire.Bombay: Bhartiya Vidya Bhawan, 1984.
Malik, Zahiruddin.	*The Reign of Muhammad shah (1719– 1748)*. Bombay: 1927.
Matto, Abdul Aziz.	*Kashmir under the Mughals (1586 – 1752)*. Srinagar: Humayun Publishing House, 1988.
Mohammed, Jigar.	*Revenue Free Land Grants in Mughal India, Awadh Region in the 17th and 18th Centuries (1658 – 1765)*. New Delhi: Manohar Publishers,2002.
Moosvi, Shirin.	*Man and Nature in Mughal Era*, symposium paper, published by IHC, 1993.
Moreland, W.H.	*From Akbar to Aurangzeb, A Study of Indian Economic History*. New Delhi: Atlantic Publishers and Distributors, 1994.
Moreland, W.H.	*India at the Death of Akbar*, An Economic Study. New Delhi: Atlantic Publishers and Distributors, 1994.
Moreland, W.H.	*The Agrarian system of Moslem India*. New Delhi: Atlantic Publishers and distributors, 1968.
Nanda, Meera.	*European Travel Accounts During the Reigns of Shahjahan and Aurangzeb* Kurukshetra: Nirmal Book Agency, 1994.

Naqvi, H.K.	*Urban Centres and Industries in Upper India (1556–1803)*, Reprint Bombay: Asia Publishing House, 1968.
Naqvi, H.K.	*Urbanisation and Urban Centres under the Great Mughals.* Shimla: Indian Institute of Advanced Studies, Dec.1972.
Newton, Dennis Kincaid.	*British social Life in India.* 1st edition Newton Abbot: Readers union,1974.
Nizami, Khaliq Ahmad.	*State and Culture in Medieval India.* New Delhi: Adam Publishers, 1985.
Ojha, P.N.	*North Indian Social Life during Mughals period.1st edition* Delhi: Oriental Publishers And Distributors, 1975.
Pant, Chandra.	*Nurjahan and her family.* Allahabad: Dandewal Publishing House, 1978.
Pant, D.	*The Commercial Policy of the Moguls.* Delhi: Idarah-i Adabiyat-i Delli,1978.
Pant, D.	*Economic History of India under the Mughals* Intro. by Dr.V.K.Saxena, Delhi: Kanishka Publishing House, 1990.
Parmu, R.K.	*A History of Muslim rule in Kashmir: 1320-1819.* Srinagar: Gulshan Books,2009.
Poole, Stanley Lane.	*Rulers of India, Aurangzeb and Decay of the Mughal Empire.*W.W.Hunter ed., Delhi: Low Price Publications,1990.
Qanungo, Kalikaranjan.	*Shershah and his times,* Calcutta: Orient Logman Ltd., 1965.
Qureshi, Ishtiaq Hussain.	*The Administration of Mughal Empire,* Reprint Delhi: Low Price Publication, 1994.
Randhawa, Mohinder Singh.	*Kangra Ragmala Paintings.* New Delhi: National Museum Delhi, 1971
Randhawa, Mohinder Singh.	*Gardens through the Ages,* Madras: The Mac Millan Company of India Ltd., 1976.

Richards, John F.	*The New Cambridge History of India*, The Mughal Empire. Reprint New Delhi: Cambridge University Press, 2011.
Rizvi, S.A.A.	*The wonder that was India* Vol.II, 1200-1700. Calcutta: Rupa&Co, 1995.
Rizvi, Saiyid Athar Abbas.	*Shah Wali Allah and His Times (A Study of 18th Century Islam Politics and Society in India).* Australia: Ma'rifat Publishing house,1980.
Sarkar, J.N.	*The Life of Mir Jumla (The General of Aurangzeb).* New Delhi: Rajesh Publications, 1979.
Sarkar, J.N.	*Mughal Administration.* 5th edition. Calcutta: M.C.Sarkar & sons, 1963.
Sarkar, Jadunath	*Fall of the Mughal empire* Vol.I, 1739–54, Reprint, Delhi: Orient Black Swan Pvt. Ltd. 1997.
Sarkar, Jagdish Narayan.	*Studies in Economic Life in Mughal India.* Delhi: Oriental Publishers and Distributors, 1975.
Sarkar, Jagdish Narayan.	*Mughal Economy:Organisation and Working.* Calcutta: Naya Prakash, Jan.1987.
Schimmel, Annemarie.	*The Empire of the Great Mughals: History, Art and Culture.* Trans. by Corrine Attwood, edited by Burzine K. Waghmar. London: Reaktion Books, 2004.
Singh, Chetan	*Region and Empire; Panjab in the 17th Century.* New Delhi: Oxford University Press, 1991.
Singh, Surinder & Ishwar Dayal Gaur ed.	*Sufism in Punjab: Mystics, Literature and Shrines.* Delhi: Aakar Books,2009.
Sinha, B.P.	*Readings in Kautilya's Arthasasthra.* New Delhi: Agam Publications,1976.
Smyth, Major G.Carmichael.	*A History of the Reigning Family of Lahore with some account of Jumoo Rajahs "The Seik soldiers and their Sirdars"* Delhi: Parampara Publications, 1979.

Srivastava, M.P.	*Policies of the Great Mughals.* Allahabad: Chugh Publications,1978.
Srivastava, M.P.	*Social Life under the Great Mughals (1526 – 1700A.D).* Allahabad:Chugh Publications. 1978.
Stocqueler, J.H.	*India Under the British Empire.* Delhi: Concept Publishing Company, 1992.
Stuart, C.M.Villiers.	*Gardens of the Great Mughals.* London: Adam and Charles Black Soho Square,1913.
Tirmzi, S.A.I	*Some Aspects of Medieval Gujarat.* New Delhi: Munshiram Manohar Lal, 1968.
Trevaskis, H.G.	*An Economic History of Punjab (From Earliest Times to 1890)* Vol.I, Gurgaon: Vintage Books,1989.
Umar, Muhammad	*Urban Culture in Northern India during the 18th Century.* Aligarh: 2001.

ARTICLES

Farooq, Abdul Aziz.	*Sodhra – History and Archaeology,* in Journal of Pakistan Historical Society, Vol.XXXVII, Part I, Karachi, Jan.1989, ed. by Dr.S.Moinul Haq.
Naqvi, H.K.	*Dyeing of Cotton Goods in Hindustan (1556 – 1803)* in Journal of Indian Textile Industry, No.VII, Ahmedabad: 1967.
Qidwai, Ikramuddin.	*Industries and Commerce in the Kingdom of Awadh* in Journal of Pakistan Historical Society, Vol.XXXVII, Part-I, Karachi, Jan.1989, Ed. by Dr.S.Moinul Haq.

JOURNALS

Journal of Indian Textile Industry No.VII, Ahmadabad, 1967.
Journal of Pakistan Historical Society, Vol.XXXVII, Part-I, Karachi, Jan.1989.

ENCYCLOPAEDIA

Bridgwater, William and Seymour Kurtz *The Columbia Encyclopaedia, 3rd Ed.* Columbia, 1968.

INDEX

A

Abdu'r Rahim, 89
Abdul Hamid Lahori, 128
Abhai Singh, 226
Ab-i-Pashan, 128
Abjosh, 30, 211
Abul Fazl, xii, 5, 8, 12, 18, 22, 29, 30, 34, 44, 45, 58, 60, 103, 107, 117, 120, 122, 125, 126, 141, 142, 143, 146, 147, 148, 149, 151, 152, 164, 167, 169, 170, 171, 190, 191, 193, 194, 195, 197, 198, 201, 204, 221, 229
achars, 167, 199
Achhabal, 64, 67, 69, 70, 97, 98, 222
Adab-i-Alamgiri, 16
Adilabad, 91
Afrasiab, 55
Afzal Khan, 130
Agra, 12, 18, 21, 22, 25, 26, 27, 28, 33, 34, 39, 40, 41, 42, 48, 55, 58, 59, 60, 62, 63, 72, 73, 75, 78, 83, 84, 89, 96, 101, 102, 103, 104, 107, 114, 132, 151, 169, 171, 174, 175, 177, 178, 199, 203, 215, 221, 222, 223, 226, 236, 238
Ahadis, 20, 170
Ahmad Shah, 40, 235
Ahmedabad, 21, 24, 94, 119, 142, 149, 163, 165, 168, 174, 175, 176, 184, 203, 242
Ahsanabad, 91, 112

Ain-i-Akbari, 8, 22, 44, 45, 117, 141, 151, 164, 195, 221, 229
Ajmer, 15, 41, 42, 63, 191, 215
Akbar, xii, 2, 4, 5, 8, 9, 12, 13, 14, 19, 20, 26, 40, 43, 62, 63, 64, 65, 73, 81, 83, 89, 90, 91, 101, 112, 126, 128, 130, 136, 139, 150, 162, 169, 170, 171, 175, 182, 184, 185, 187, 189, 200, 201, 213, 215, 221, 222, 225, 229, 232, 236, 239
Akbarabad, 39, 85, 90
Akhbarat, 40, 51, 235
Alampur, 227
Alauddin Khalji, 2
Alfonso, 16, 19
Ali Mardan Khan, 76, 79, 82, 88, 93, 113
Ali Muhammad Khan, 21, 119
Alivardi Khan, 95
Allahabad, 27, 36, 41, 42, 73, 124, 152, 169, 188, 208, 216, 240, 242
almonds, 10, 18, 19, 28, 33, 34, 36, 40, 127, 130, 131, 141, 142, 167, 177, 178, 223
alu-balu, 12, 211
aluchas, 30, 141
Amal-i-Saleh, 18, 46, 47, 48, 49, 50, 184
Amber, 202, 215, 225, 226
amber-i-ashhab, 195
ambli, 6, 155
Amin Ahamd Razi, 90
Amir Khusrau, 4, 6
Amir Qazwin, 56

amla, 2, 7, *141, 155, 192*
Amrd-fal, 7
Ananas, 11, 18, 218
anardana, 179, 192
anbali, 7, *140*
Anguri bagh, 83, 95, 215
Anguri Bagh, 88
Anhilwara, 38
apples, 3, 4, 5, 8, 11, 14, 16, 18, 20, 21, 30, 33, 34, 35, 36, 37, 38, 40, 42, 131, 141, 142, 167, 177, 178, 190, 193, 194, 222, 223
apricot, 2, 7, 10, 14, 30, 37, 70, 141
Arabia, 175
Aram Bagh, 58, 215
araq-i-Chameli, 164
Araq-i-Sewti, 164
Arayankas, 115
argajah, 133
Argajah, 199
Arjuand Banu *Begum*, 84
Arra-kash, 171, 211
Arthashastra, 164, 200
Asad Khan, 132
Asaf Khan, 15, 23, 91, 92, 99, 239
Asmat Begam, 164, 195
asoka, xi, *1, 115, 122, 125*
Assam, 18, 41, 146, 147, 162, 163, 168
atimukta, *1*
Attock, 26, 152
Aulugh Beg, 56
Aurangabad, 19, 25, 41, 87, 112, 173, 216
Aurangzeb, 4, 9, 15, 16, 20, 24, 28, 36, 44, 46, 48, 49, 50, 68, 79, 80, 81, 83, 86, 87, 88, 93, 103, 108, 111, 112, 121, 129, 130, 131, 132, 137, 138, 139, 172, 183, 184, 188, 206, 207, 231, 233, 237, 239, 240, 241
aurta bagh, 56

Awadh, 5, 11, 17, 28, 39, 41, 42, 50, 116, 136, 170, 203, 235, 236, 239, 242
Azad Bilgrami, 105
Azamgarh, 170

B

Baba Jiwan Shah, 117
Baba Latifuddin Rishi, 117
Babul, 144, 149, 166, 197, 217, 218
Babur, 4, 5, 6, 7, 9, 12, 21, 22, 43, 44, 55, 56, 57, 58, 59, 60, 62, 88, 96, 98, 99, 100, 101, 102, 103, 107, 114, 120, 130, 133, 140, 143, 146, 149, 158, 160, 164, 180, 191, 192, 193, 196, 197, 198, 215, 221, 230, 237, 238
Baburnama, *43, 44, 45, 47, 51, 55, 56, 102, 106, 107, 111, 136, 180, 181, 183, 184, 206, 207, 208, 230, 237*
Babylon, 52
Badakshan, 12, 13, 14, 20, 177, 221
Baden-Powell, 153
bagh, 56, 57, 60, 64, 65, 66, 68, 69, 71, 73, 74, 75, 77, 79, 80, 81, 83, 85, 87, 90, 92, 94, 95, 96, 98, 99, 101, 102, 103, 120, 221, 223
Bagh -i-Safa, 60
Bagh -i-Saifabad, 92
Baghalpur, 40
Bagh-i-Afzalabad, 92
Bagh-i-baharara, 71
Bagh-i-bihisht, 56
Bagh-i-bulandi, 55
Bagh-i-chanar, 56
Bagh-i-Dilkusha, 55, *81*
Bagh-i-diwan khana, *100*
Bagh-i-Gulafshan, 59
Bagh-i-Husnabad, 92
bagh-i-Jalankhana, 56

bagh-i-mahtab, 56
Bagh-i-Maidan, 56
Bagh-i-Murad, 92
Bagh-i-Nasim, 92
Bagh-i-NurAfzai, 92
Bagh-i-Shamel, 56
Bagh-i-wafa, 56
Bagh-i-zafran, *101*
Bagh-i-Zarafshan, 59
Baglana, 18, 38, 41, 139
Bagpur, 39
Bahadur Shah, 25, 88, 178
Bahadur Shah Jafar, 88, 128
bahar-i-naranj, *194*
Bahauddin Sahib, 158
Bahist bagh, 95
Baihraich, 179
Baijnath, 159
Balasore, 153
Balkh, 12, 36, 131, 177, 193
Balsar, 203
bamboo, 118, 120, 158, 166, 167, 170, 204
banafsha, 24
banana, 1, 12, 21, 127
Band Salora, 54
Banyan tree, 123, 124
baolis, *57*, *117*, *211*
Bara Palah bridge, 27
baradari, 95
Barbosa, 134, 155, 165, 176, 199
Bargad tree, 124
Basant Panchami, 129
Batala, 129, 171
Bayana, 33, 38, 41, 107
Beas, 150
bed-i-Mushk, *155*
Begam Mukhtar Mahal, 95
Begum ka bagh, *79*, *102*
Begumabad, *69*
Benaras, 170, 178, 203

Bengal, 7, 16, 17, 18, 21, 25, 26, 34, 36, 37, 38, 40, 41, 42, 87, 118, 146, 147, 151, 152, 153, 157, 158, 159, 162, 163, 165, 167, 168, 169, 170, 171, 172, 173, 175, 176, 191, 192, 194, 203, 204
Berar, 41, 112
Bernier, Francois, 5, 16, 18, 25, 28, 36, 37, 41, 44, 46, 48, 49, 50, 51, 68, 70, 78, 81, 85, 87, 108, 109, 111, 119, 132, 133, 136, 139, 141, 147, 149, 150, 157, 158, 167, 169, 171, 177, 180, 181, 183, 184, 185, 186, 187, 189, 191, 193, 204, 206, 207, 209, 222, 230, 231
bers, *8*, *17*, *141*
betel leaf, 39, 129
Bhakkar, 60
Bharatpur, *215*, *226*
Bhimber, 133, 174
Bhira, 60
Bhoj pattra, *159*
bholsari, *164*
Bibi Mubarika, 60
Bihar, 41, 42, 47, 118, 130, 147, 162, 163
Bijapur, 133, 147
Bijbehara, 77
Bikaner, 225
black berries, 39, 132
Bokhara, 36, 141, 177
Bonford, 176
Borneo, 175
Brazil, 19
Britain, 96
British, 18, 25, 46, 48, 51, 81, 96, 113, 136, 137, 139, 166, 180, 182, 183, 184, 185, 186, 208, 236, 240, 242
Britishers, 40, 96, 199
Bukhara, 12, 131, 177, 221

Bundelkhand, 41
Bundi, 227
Burhanpur, 15, 35, 39, 40, 42, 89, 103, 132
Bussara, 177
buyutat, 169

C

Cambay, 118, 178
Camphor, 147
Cape Comorin, 175
Careri.Abbe, 160, 183, 184, 198, 207, 231
Cashew-nut, xii, 18
Central Asia, 4, 173, 175, 177
Ceylon, 151
Chahar Chinar, 71
Chamba, 39, 153, 174, 227
Chambal, 160
chambeli, 22, 23, 25, 121, 133, 146, 164, 190, 194, 196, 202
Champa, xii, 23, 25, 144, 220
champaka, xi, 115, 125
Champaner, 165, 203
Chanaram Chakravarty, 199
Chandan, 143, 148
Chander Gupta Maurya, 53
Chandra Bhan Brahman, 49, 81, 110
char bagh, 56, 67
Chashma-i-Shahi, 76, 77
Chaukandi, 60, 107
Chenab, 93, 153, 174
cherries, 8, 14, 30, 34, 35, 36, 41, 42, 141, 222
Chib tribe, 133
chiku, 19
China, 65, 147, 148, 175, 177
china palu, 163
chinar, 13, 65, 67, 71, 92, 121, 222
Chishti, 87
Chitor, 25

chutneys, 167, 199
chuwah, 133, 165
Citron, 7
coconut, 6, 7, 10, 32, 37, 38, 40, 120, 121, 127, 134, 141, 142, 154, 155, 156, 173, 174, 178, 194, 199, 201, 213, 219, 222
Congo, 177
Cooch Bihar, 39
coquinhos, 155
coral, xi, 115, 132
Coromandel, 151, 175
cypresses, 11, 15, 91, 221, 222, 223

D

Dacca, 170
daffodils, 15, 223
Daftar yu Sunluq, 212
Dak Chawki, 21
Dal lake, 24, 28, 64, 65, 71, 77, 98, 121, 134, 222
Daman, 19
Damascus, 178
Damishqi, 173, 178
Dara Shikoh, 77, 84, 128
Daroga Bayutat, 71
date palm, 6, 7, 120, 160, 197
dates, 8, 10, 33, 34, 40, 132, 141, 142, 177, 178
Daulat Khan, 191
De Laet, 26, 48, 49, 85, 111, 121, 137, 158, 184, 186, 231
Deccan, 15, 28, 38, 39, 40, 41, 42, 87, 112, 130, 132, 134, 163, 235
Deeg, 215, 226
Dehra bagh, 58, 63, 73
Delhi, 2, 5, 6, 14, 21, 25, 27, 28, 36, 37, 39, 41, 42, 43, 44, 45, 46, 47, 48, 49, 53, 54, 55, 62, 63, 75, 78, 79, 80, 81, 86, 88, 95, 98, 102, 103, 104, 106, 107, 108, 109,

110, 111, 113, 114, 115, 128, 130, 132, 133, 136, 137, 138, 139, 141, 150, 169, 177, 181, 182, 183, 185, 188, 206, 215, 223, 225, 227, 229, 230, 231, 232, 233, 234, 235, 236, 237, 238, 239, 240, 241, 242

Della Valle, 10, 19, 23, 45, 46, 47, 49, 123, 125, 137, 159, 180, 183, 192, 206, 231

Dennis Kincaid, 96, 113

desi, 163

Dharam Mangal, 199

Dholpur, 54, 90, 103

dhup, 200

Dinpanah, 62

Dipalpur, 12, 89

Doab, 17, 170

doondah, 153

Dutch, 72, 74, 166, 169, 175, 176, 234

E

East India Company, 153

Edward Terry, 10, 34, 121, 155, 158, 160, 183, 184

Egyptians, 52

Europe, 10, 16, 19, 24, 34, 37, 153, 167, 174, 176

F

Faiz Baksh, 75, 103

Faizabad, 5, 11

Farah-Baksh, 68

Farghana, 55, 177

farmans, 26, 167

Farrukhabad, 18, 95, 113

Farruksiyar, 80, 88

Fatehgarh, 18

Fath bagh, 90

Father Monserrate, 3, 26, 28, 30, 43, 49, 51, 62, 90, 108, 112, 119, 132, 136, 139, 150, 181, 182, 201, 208, 232

Fergusson, 115, 136

Fidai Khan, 93

figs, 3, 8, 10, 11, 34, 35, 37, 38, 40, 42, 115, 116, 127, 132, 141, 142, 168, 177, 223

Firdausi, 55

Firoz Tugluq, xi, 2, 3, 54

French, 5, 16, 17, 24, 63, 96, 150, 234

Friar Oderic, 153

Fyzabad, 95

G

galgal, 7

Gandak, 39

Gandha Baniks, 203

Ganga, 39, 175

Gautam, 122

Gesu Daraz, 91, 122

Ghazipur, 164, 174

Ghoraghat, 163

Gilgit, 162, 163

Goa, 16, 18, 19, 34, 36, 37, 41, 46, 90, 119, 134, 142, 178, 179

Golconda, 36, 41

Golkunda, 160, 172, 204

Gosains, 170

grapes, 2, 4, 6, 10, 11, 12, 17, 19, 20, 28, 29, 30, 31, 32, 33, 34, 35, 36, 37, 38, 39, 40, 41, 42, 121, 130, 131, 132, 141, 142, 161, 177, 179, 193, 198, 199, 201, 223

guavas, 8, 14, 17, 20, 29, 39, 40, 141, 142

Gujarat, 7, 14, 18, 34, 36, 38, 40, 41, 42, 54, 55, 90, 120, 134, 146, 147, 150, 154, 155, 165, 170, 174, 175, 178, 199, 242

Gulafshan garden, 18

Gulam Hussain Zaidpuri, 5, 17, 208

gular, 6, 8, 141

Gulbadan Begam, 4, 44, 56, 59, 60, 107
Gulbarga, 90
gul-i-Aftab, 146
Gul-i-karna, 196
gul-i-kawal, 146
gul-i-surkh, 146, 194
Gul-i-zafaran, 121
Gulkama, 194
gum lac, 147, 173
gur, 167, 172
Gwalior, 22, 50, 54, 57, 194, 203

H

Haft Iqlim, 51, 90, 112, 234
Hajipur, 39, 41
hamamas, 133
harapalu, 163
Harappan, xi, 122
Hari Parbat, 64, 65
Hasan Abdal, 73
Hasanpur, 39
Hasilpur, 40
Hayat Baksh, 78
Hazrat Zainuddin Rishi, 117
henna, 161, 164, 190
Himalayas, 225, 227
Hindal, 61, 101, 103
Hindustan, 3, 4, 6, 7, 8, 12, 13, 15, 21, 24, 29, 30, 43, 57, 67, 106, 146, 147, 149, 167, 168, 184, 191, 192, 193, 196, 235, 242
Hisar Firoza, 3, 41
Holi, 128
Holland, 175
Hugli, 41
Humayun, 27, 45, 48, 60, 61, 62, 89, 98, 101, 102, 103, 107, 108, 111, 114, 201, 215, 232, 239
Humayun's tomb, 27, 62, 98, 103

I

I'timad-ud-Daula, 73, 215
Ibaratnama, 151
Ibrahim Adil Shah, 91, 122
id-i-gulabi, 128
Id-i-Milad, 102
Inayat Khan, 75, 76, 80, 81, 82, 84, 85, 102, 128, 133, 150, 151
India, 1, 2, 3, 4, 5, 6, 7, 9, 10, 11, 12, 13, 15, 16, 17, 18, 19, 21, 22, 23, 25, 26, 27, 28, 29, 30, 31, 33, 34, 35, 36, 37, 39, 40, 41, 42, 43, 44, 45, 46, 47, 48, 49, 50, 51, 53, 54, 55, 57, 58, 59, 61, 62, 63, 64, 65, 66, 67, 69, 71, 73, 74, 75, 77, 79, 81, 83, 85, 87, 89, 91, 93, 94, 95, 96, 97, 98, 99, 101, 102, 103, 105, 106, 107, 109, 111, 113, 114, 115, 117, 119, 120, 121, 122, 123, 125, 127, 129, 131, 133, 135, 136, 137, 138, 139, 140, 141, 142, 143, 145, 146, 147, 148, 149, 151, 153, 155, 157, 158, 159, 161, 162, 163, 164, 165, 167, 169, 170, 171, 173, 174, 175, 176, 177, 178, 179, 180, 181, 182, 183, 184, 185, 186, 187, 188, 189, 190, 191, 193, 195, 197, 199, 200, 201, 203, 205, 207, 208, 209, 213, 218, 219, 222, 223, 225, 227, 229, 230, 231, 232, 233, 234, 235, 236, 237, 238, 239, 240, 241, 242, 243
Indian Ocean, 153
Indica, 1, 53, 223
indigo, 161, 162, 167, 176
Indus, 25, 26, 109, 153, 176
Iran, xii, 5, 12, 25, 130, 170, 173, 191
Iraq, 175
Islam Shah, 62
Italy, 30, 34

itr, 24, 164, 174, 195, 200, 203, 205
itr-i-Jahangir, 24

J

jack fruit, 3, 6, 7, 8, 18, 34
Jagmandir, 227
Jagniwas, 227
Jagra, 160
Jahanara, 67, 79, 80, 92
Jahandarshah, 88
Jahangir, xii, 5, 9, 10, 12, 13, 14, 15, 18, 19, 20, 21, 23, 24, 26, 27, 33, 34, 35, 42, 45, 49, 59, 63, 64, 65, 66, 67, 68, 69, 70, 71, 72, 73, 74, 75, 83, 90, 91, 104, 118, 124, 127, 130, 131, 134, 136, 138, 139, 141, 146, 149, 150, 152, 155, 162, 169, 175, 191, 192,194, 195, 200, 221, 222, 223, 232, 234
Jahangiri itr, 164
Jai Singh II, 225
Jaipur, 47, 89, 215, 225, 232
Jalalabad, 41
Jalandhar, 72
Jalesar, 164, 174
jamans, 33, 140
jambol, 3
Jammu, 106, 117, 130, 136, 229, 238
Jamuna, 4, 9, 39, 58, 59, 61, 78, 90, 171, 175
Japan, 175
Jarogha bagh, 66
jasmine, 1, 22, 23, 24, 25, 39, 40, 59, 61, 120, 126, 146, 147, 164, 190, 194, 203, 223, 226
jasun, 21, 120, 146
Jaswant Singh, 86, 132
Jaunpur, 28, 164, 174, 203
Java, 146, 147, 175
Jehlum, 77, 92, 153
Jellall tribe, 133

Jodhpur, 3, 25, 36, 54, 225, 226

K

Kabul, 4, 7, 8, 9, 12, 13, 14, 20, 26, 28, 30, 33, 34, 35, 41, 42, 55, 56, 57, 60, 66, 67, 75, 84, 89, 101, 103, 141, 150, 169, 174, 177, 192, 193, 215, 223
kacchis, 17, 117
kachnar, 1
kadali, 1
kadals, 150
Kadamba, 1, *115*, *125*, *212*
Kairana, 14, 15, 91, 191
Kakrali, 89
Kalanbak, *196*, *202*
Kaldakahar, 60
Kalidasa, 53
Kalka, 93, 215
Kalpi, 41
Kama Deva, 125
Kamasutra, 53
Kambayet, 178
Kamran, 61, 81, 102, 103
Kamroop, 203
Kamyak, 115
Kandhar, 33, 223
kaner, 22, *121*
Kangra, 30, 227, 228, 240
Kanjak wood, 143
Kannuj, 62
Kanpur, 17
kanwal, 24, 25
Kariz, 20, 131, 192
karkhanas, *169*, *171*, *173*, *211*
Karnal, 80
karunda, 6, 7, *141*
kasa-i-ghichaks, 156
Kashmir, 2, 3, 8, 9, 13, 14, 15, 20, 22, 23, 24, 25, 28, 29, 30, 33, 35, 36, 37, 39, 41, 42, 45, 48, 49, 60,

64, 65, 66, 67, 68, 69, 70, 71, 72, 73, 75, 76, 77, 78, 80, 83, 87, 89, 92, 96, 97, 98, 99, 100, 103, 108, 109, 110, 117, 118, 119, 120, 121, 133, 134, 137, 141, 147, 148, 149, 150, 151, 152, 153, 157, 158, 159, 162, 163, 166, 171, 172, 173, 174, 182, 183, 192, 195, 198, 202, 207, 215, 222, 224, 229, 231, 237, 238, 239, 240
kathal, 3, 38
Kautilaya, 164
Kerala, 175
ketki, 22, 23, 25, 121, 190, 194
Khajur, 122
Khalifa, 89
Khandesh, 25, 36, 40, 41, 42, 112
Khan-i-Khana, 89, 90
kharbuza, 38, 39, 134
Kharbuza-i-karez, 131
Khas, 166, 212
Khatam bandi, 158
Khavaja Kilan, 57
khayaban, 25, 169
khilat-i-khasa, 132
Khulasat ut Tawarikh, 17
Khurasan, 15, 29, 36, 192
khurma-i-Hind, 120
Khurram, 66, 69
Khusaru, 73, 74
khushbu khana, 201
Khusrau Bagh, 216
Khwaja Dost Munshi, 61, 103
Khwaja Ghazi, 61
Khwaja Kilan, 9
Khwanda Mir, 61, 101
Kimsuka, 1
kishmish, 30, 131, 141
kishmishi, 33, 34, 42, 141
Kishtwar, 166, 238
kiura, 21, 22, 146

koeris, 17, 117
Koh-i-daman, *193*
Kolaras, 50, 142
Konkan, 41
Kotah, 41
Kuh-i-maran, 64

L

Lac, 167, 174, 176
Lahari-Bandar, 176
Lahore, 12, 16, 17, 21, 23, 25, 26, 27, 33, 35, 36, 38, 41, 42, 60, 73, 74, 75, 80, 81, 82, 83, 87, 89, 90, 92, 93, 96, 98, 99, 102, 103, 104, 110, 113, 138, 139, 146, 151, 152, 168, 169, 171, 174, 176, 177, 178, 203, 215, 223, 227, 238, 241
Lake Manasbal, 66
lake Pichola, 226
Lakhira, *170*, *171*, *212*
Lakshmi, 122, 125
Lal Bagh, 89
lemon, 7, 37, 141, 162, 179, 192, 193, 199, 204, 227
lichi, 19
lignum aloes, 133, 195
limes, 3, 40, 142, 167, 193, 204
Linschoten, 18, 163
Lodi, 54, 57, 191
lokat, 19
London, 43, 45, 106, 107, 108, 136, 153, 154, 177, 230, 232, 233, 235, 236, 241, 242
lote fruit, 6, 7, 141
Lucknow, 17, 95
Ludhiana, 150
Luk, 166

M

Machchi Bhavan, 83
Machivara, 150
madad-i-maash, 116, *134*
Madin Sahib, 158
Madras, 25, 43, 96, 150, 151, 182, 239, 240
Magasari, 148
Mahabharata, xi, *53*, *115*
Maharashtra, 18, 37, 46, 238
Mahtab Bagh, *78*, *223*
mahuwa, 6, 7, *120*, *121*, *140*, *141*, *143*, *159*, *193*, *194*, *197*
Makhduma-i-Jahan, *91*
Malabar, 134, 147, 151, 174
Malacca, 16, 178
Maldah, 163, 172
Maldives, 178
Malik Haider, 71
Malik Muhammad Jaisi, 4, 164
Malika-i-Zamani, 80
malis, *17*, *117*
Malwa, 25, 36, 42, 147
Malwah, 40, 41
Man singhi paper, *172*
mandali, *165*
Mandelslo, 159
Mandor, 226
Mandu, 25, 130, 216
mango, xi, *1*, *3*, *5*, *6*, *11*, *15*, *16*, *17*, *19*, *35*, *36*, *38*, *39*, *40*, *41*, *81*, *96*, *115*, *116*, *118*, *119*, *120*, *122*, *125*, *133*, *141*, *157*, *167*, *173*, *191*, *193*, *199*, *219*, *222*
Manrique, 111, 194, 203
mansabdars, *20*, *86*, *170*
Manucci, 5, 19, 24, 27, 34, 35, 44, 45, 46, 47, 48, 49, 79, 84, 87, 108, 110, 111, 125, 127, 131, 132, 137, 138, 139, 160, 164, 167, 168, 169, 173, 176, 182, 183, 184, 186, 187, 188, 189, 191, 192, 193, 194, 195, 196, 197, 200, 201, 203, 206, 207, 208, 233
Marcopolo, 178
marigold, 3
Marwari, 118
Massir-i-Alamgiri, *86*
Masudat Munshi, 16
Masulipatam, 150
Maulvi Khairuddin Lahori, 151
Mauryan, 53
Megathenes, 53
Meghaduta, *53*
melons, 3, 4, 5, 6, 8, 10, 11, 12, 13, 14, 15, 16, 17, 18, 19, 20, 29, 30, 31, 33, 34, 35, 36, 37, 38, 39, 40, 41, 121, 130, 131, 132, 141, 142, 177, 178, 192, 199, 222
Mesopotamia, 52
Mewah Khana, 20
Mewa-i-Khuskhi, *30*
Mewa-i-Shirin Hindi, *30*, *34*
Mewa-i-tars, *30*, *34*, *212*
Mewar, 106, 225
Mickapur, 194, 203
Middle East, 175
Mighush, *30*, *34*, *212*
Mir Dard, 105
Mirat-i-Ahmadi, *21*, *50*, *182*
Mirza Haider Dughlat, 8, 89, 121, 158, 162
Mirza Shah Hussain Samandar, 60
Misah-yi-sayila, *147*
Misah-yi-yabisa, *147*
mogra, *25*
Mohammad Amin Qazvini, 29, 90, 112
Mohammed Khan Bangash, 94
Mohammed Shah Rangila, 80
Mohibul Hasan, 99, 107, 113

monkey jack, 6, 141
mosseri, *164*, *195*
Moti Masjid, 86
motia, *25*
Motibagh, 73, 95
Muga, *163*
Mughals, xi, xii, xiii, xiv, 5, 9, 17, 18, 24, 39, 42, 44, 45, 46, 48, 54, 55, 64, 83, 88, 92, 95, 96, 97, 98, 100, 102, 103, 104, 106, 107, 108, 109, 110, 111, 114, 116, 140, 153, 173, 184, 186, 188, 189, 191, 201, 208, 221, 222, 223, 227, 230, 236, 237, 238, 239, 240, 241, 242
Muhammad Rida, 15
Muhammad Shah, 11, 17, 21, 28, 39
Muhammed Riza Khan, 95
Muhammed Saleh Kamboh, 18
Mui'nu-d-din Sanjari chishti, 74
Mukundram, 199
mulberries, *8*, *11*, *14*, *35*, *141*, *163*
mulberry, 8, 14, 26, 28, 34, 40, 121, 149, 162, 163, 222, 223
mulsari, *23*
Multan, 26, 27, 41, 42, 152, 168, 169, 174, 176, 177
Mumtaz Mahal, 78, 103
munaqqas, *178*
Muqarrab Khan, 13, 14, 15, 91, 191
Murad Bakhsh, 86
murrabba, *167*
Murshid Quli khan, 95
Murshidabad, 95, 170, 172
Muscat, 177
myrobalans, 34, 37, 167, 176, 192, 204

N

Nadir Shah, 81, 95
nag- champa, xi, *115*
nag-kesar, *146*

Nakodar, 74
Nandurba, 18
NaqshiI-jahan, *56*
narang, 3
nargis, *22*, *146*, *164*
Naseem bagh, 64, 65, 222
nashpati, *14*, *16*, *29*
Naulakh bagh, *81*
Nauroz, *102*
Nausari, 165, 203
Navsari, 122
Nawab Shuja-ud-daula, 5, 95
Nazir Ahmad, 95
neem, *80*, *223*
Nicholas Withington, 90
Nicole Afonco, 16
Nicolo de conti, 161
Nishat, *69*, *92*, *95*, *98*, *99*, *215*
Nishkar, *38*, *131*, *212*
nistari, *163*
Nizammu'd-din Ahmad, 90
Noor *bagh*, 71
Nooruddin, 47, 50, 90
Nurjahan, 15, 45, 47, 66, 67, 68, 69, 70, 71, 72, 73, 91, 103, 109, 112, 113, 130, 164, 165, 180, 195, 196, 200, 207, 237, 240
Nur-manzil garden, 23, *72*

O

oak, 121, 153
oleander, xi, 21, 115, 146
Omrahs, *28*, *133*
oranges, 2, 3, 7, 8, 10, 17, 29, 34, 37, 38, 39, 40, 41, 56, 120, 132, 133, 141, 142, 176, 179, 190, 192, 199, 200, 221, 223
Orissa, 24, 47, 118, 146, 147, 152, 158, 159, 163, 174
Ovington, 159
Oxus, 176

P

Padshannama, 56
Pahar Khan, 12
Paien bagh, 83
Pakhli, 39
palaeolithic, 115
Palmeira brava, 154
Pan, 87, 212, 216
Panipat, 100
paniyala, 6, 7
Pankhon ka mela, 102
papaya, xii, 19, 18, 42, 142
Pari Mahal, 77, 215
Parindas, 153
Parsis, *121, 132, 204*
Parvati, 123
Parvez, 71, 72
Patal, 145, 166, 171
Patna, 26, 35, 39, 47, 159, 175, 230, 236, 239
Patoles, 175, 212
Pattan, 38, 123
paywandi, 14
peaches, 11, 16, 18, 20, 28, 30, 34, 35, 37, 38, 39, 42, 87, 141, 177, 190, 193, 223
pears, 5, 8, 11, 14, 16, 30, 33, 34, 35, 36, 37, 131, 141, 142, 177, 222
Pegu, 175
Pelsaert, 33, 49, 72, 73, 74, 109, 119, 127, 133, 136, 138, 142, 149, 168, 174, 176, 177, 178, 180, 181, 183, 186, 187, 188, 189, 195, 207, 208, 234
Perfumes, xii, 133, 178, 199, 200
Persia, 3, 13, 17, 28, 33, 36, 38, 55, 97, 123, 131, 141, 161, 163, 166, 173, 175, 176, 177, 178, 193, 201, 234
Peshawar, 177
Peter Mundy, 10, 26, 35, 63, 73, 134, 141, 159, 221, 223
phalsa, 7, *141*
Philippines, 175

p
pine-apple, xii, 18, 23, 34, 142, 155, 157, 218, 222
pinjara saz, 170
Pinjaur, 93, 94, 215
pipal, xi, *115, 121, 122, 123, 125, 126, 137, 159, 223*
Pir Buddhan Ali Shah, 117
Pir Lakhdatta, 117
Pir Mitha, 117
Pir Panjal, 174
Pir Roshan Wali, 117
pistachios, *28, 33, 36, 130, 131, 142, 178*
piyadah, 20
plantains, 6, 8, 12, 34, 37, 40, 41, 120, 130, 141, 142, 221
plum, 2, 7, 12, 40, 70, 155, 167, 211, 217, 224, 227
plums, 8, 10, 16, 36, 37, 38, 40, 130, 131, 141, 142, 177
pomegranates, 3, 4, 8, 10, 14, 17, 18, 20, 28, 30, 33, 34, 36, 38, 40, 41, 56, 131, 141, 142, 162, 177, 192, 199, 221, 223
Pondicherry, 17, 96
Portugal, 179, 201
Portuguese, 16, 18, 19, 46, 90, 134, 175, 178, 197, 211
Pravarasena, 68
Pumple-nose, 219
Punjab, 39, 48, 117, 136, 168, 176, 178, 182, 227, 235, 237, 241, 242

Q

qalams, 166
Qannauj, 164, 203
Qanun-i-Humayuni, 61, 108, 114, 233

qara qashuq, 156
Qasim bazaar, 163, 169
qaysuri, 148
quince, 7, 33, 34, 223
Qurqury, 148

R

Rabia Darauni, 216
Raj nagar, 227
Raja Govardhan Chand, 227
Raja Man Singh, 225
Rajauri, 108, 215
Rajmahal, 40
Rajputana, 41, 54, 226
Rajputs, 166, 225
Ralph Fitch, 90, 118
Ramayan, 53
Rana Jagat Singh, 227
Rao Chattar Sal, 227
ratanmala, 22, 146, 161, 162
Ravi, 82, 113, 150, 153, 171
Rawalpindi, 73, 109
raybel, 146
Red Fort, 78, 86, 110, 215
Rehla, 3
Rihabi, 148
rihan, 24
rishis, 1, *117, 118*
Rohilkhand, 170
rose, 1, 12, 22, 24, 25, 57, 58, 68, 72, 104, 126, 128, 130, 131, 133, 138, 155, 164, 165, 173, 174, 178, 194, 195, 196, 199, 200, 203, 211, 220
Roshanara, 79, 215
Ruh-afza, 194

S

Sa'adat khan, 95
Sabnak, 153
sada-fal, 7
Sadiq Khan, 90

Safdar Jang, 17
saffron, 64, 126, 133, 134, 147, 155, 161, 164, 165, 166, 174, 178, 190, 199
Saharanpur, 91
Sahelion-ki-bari, 98, 216
sahibi, 4, *33, 34, 42, 141, 193*
sal, xi, 1, *115, 120, 122,* 144, *171, 174,* 227
Salabat Khan, 90, 94
Samarkand, 4
San hemp, 166
sandal, 121, 125, 126, 127, 129, 130, 133, 140, 146, 147, 155, 164, 165, 173, 174, 175, 176, 177, 190, 194, 195, 199, 200, 222
Sangtara, 6, 7, 16
Sansar Chand, 227
Santak, 133, 165, 194, *199*
*Sar-i-darakhti,*xii, 19
Sarkar Mungir, 159
Sarkhej, 90, 122
Sarsuti, 123
Satgaon, 41, 176
Saurat bagh, 56
serais, 62, 72, 80, 90
sericulture, 162, 218
sewti, 146, 164, 190
Shadoof, 53, 106
Shah Abbas, 130
Shah Gulam Badshah, 117
Shah Shuja, 40
shahalu, 13, 141
Shahdara, 73, 88, 113, 215
Shah-i-Hamadan, 158
Shahjahan, 5, 9, 11, 15, 16, 24, 35, 50, 51, 55, 65, 66, 68, 69, 71, 75, 76, 77, 78, 79, 80, 81, 82, 83, 84, 87, 92, 96, 98, 102, 103, 104, 109, 110, 111, 112, 114, 130, 138,

139, 182, 184, 215, 222, 223, 227, 233, 239
Shahjahanabad, 39, 78, 79, 80, 81, 88, 96, 102, 110, 129
Shahrara garden, 26
Shah-toot, 213
Shaikh Mansur, 201
shakhtalu, 16
shalamar bagh, 66
Shalamar garden, 75, 93, 98
Sharifa, 213
shatranji, 171
Sheikh Salim, 62
Shershah Suri, 62
Shikaras, 153
Shiva, 125
Shivaji, 18, 46, 50, 238
Sialkot, 172
Sikandar Lodi, 3, 57
Sikandra, 63
Sikri, 59, 60, 62, 63, 145, 167, 215, 221
silk, 8, 68, 125, 127, 140, 162, 163, 164, 167, 168, 169, 170, 172, 173, 174, 175, 176, 177, 178, 179, 212, 217
Sindh, 41, 71, 153, 176, 177, 178
singhara, 34, 141
Sir Henry Middleton, 176
Sir Thomas Roe, 66, 130, 232
Sirhind, 11, 14, 74, 79, 85, 90, 102, 111, 119, 134
sirisa, 1
Sironj, 203
Sisso wood, 153
Sitaphal, 226
Sodhra, 93, 112, 150, 242
Somnath, 123
Sonargoan, 25
Spanish-wax, 166
Sri Krishna, 125
Srinagar, 45, 64, 65, 68, 73, 76, 77, 150, 182, 239, 240

Streynsham Master, 150
Sufaida, 17
sugandh gugala, 165, 195
sugar, 38, 127, 129, 131, 160, 161, 167, 168, 173, 175, 176, 177, 191, 192, 194, 197, 198, 199
sugarcane, 8, 12, 37, 38, 41, 121, 125, 141, 167, 173, 190, 197, 221
Sujan Rai Bhandari, 16, 78, 93, 129
Sultan Mahmud, 54, 120
Sultanpur, 72
Sumatra, 146, 175
Surant, 18
Surat, 18, 19, 25, 34, 38, 46, 67, 121, 122, 123, 132, 134, 147, 163, 165, 166, 168, 172, 173, 176, 203
Sutlej, 150
Sweet Cherry, 14
sygwan, 146, 151

T

tabdan tarashi, 158
Tabqat-i-Akbari, 54, 89, 107, 111, 112, 136, 189, 229
Tabqat-i-Baburi, 56, 59, 106, 107, 181, 183, 206, 233
Taj Mahal, 75, 84, 85, 215
Takht-i-rawans, 153
talipot, 120
tar, 7, 40, 120, 141, 157
tarbuz, 38
Tari, 159, 160, 198, 204, 218
Tarikh-i-Akbari, 62, 108, 138, 234
tarkuls, 8
Taryag-i-zahar, 154
Tasbih gulal, 193
tassar', 163
Tavernier, 27, 37, 50, 87, 111, 124, 132, 137, 139, 142, 147, 160, 162, 163, 165, 168, 172, 180, 181,

184, 185, 186, 187, 188, 189, 196, 198, 204, 207, 209, 235
Tavi, 93, 113
teak, 120, 146, 151, 174, 217
Terry, 10, 45, 119, 122, 129, 160, 191, 198, 204, 235
Thaneswar, 175, 203
Thatta, 36, 38, 41, 173, 174, 176, 177, 178
The ShahJahan-nama, 41
Thevenot, 19, 168, 231
Thomas Coryat, 26
Thomas Moore, 73
Tibet, 162, 163
Timur Beg, 55
Tira Sujanpur, 227
Trihut, 40, 41
Tulips, 22
Turan, xii, 5, 12, 25, 141, 170, 191, 212
turanjes, 38
Turkestan, 13, 17, 38
tur-khana, 60
Tuzuk-i-Jahangiri, 45, 64, 136, 232

U

Udaipur, 98, 216, 226
Udyan, 53, 213
Ujjain, 86
Uttar Pradesh, 95

V

Valentina, 120
Vatsayayana, 53
Vedic, xi, *1*
Verinag, 66, 67
Vishnu, 122

W

W.Hastings, 178
Wah *bagh*, 64, 73, 215
water-melons, 10, 37, 38, 39, 141
Wazirabad, 112, 152, 153, 174
William Finch, 11, 23, 26, 27, 34, 47, 48, 74, 89, 92, 112, 141, 147, 160, 169, 174, 180, 181, 187, 223, 232

Y

Yazd, 20

Z

zabad, *164*, *196*
Zabit Khan, 91
Zablistan, 20
Zafar khan, 77
Zafarbad, 77
Zahara, 58, 59
Zahara bagh, 58, *59*
Zain Khan, 56, 181, 183, 184, 206
Zain-ul Abidin, 3
Zebunissa, 80, 87
Ziauddin Barni, xi, 3
ziba, *24*
zoruk, *153*

BARADARI AT SHALAMAR GARDEN KASHMIR

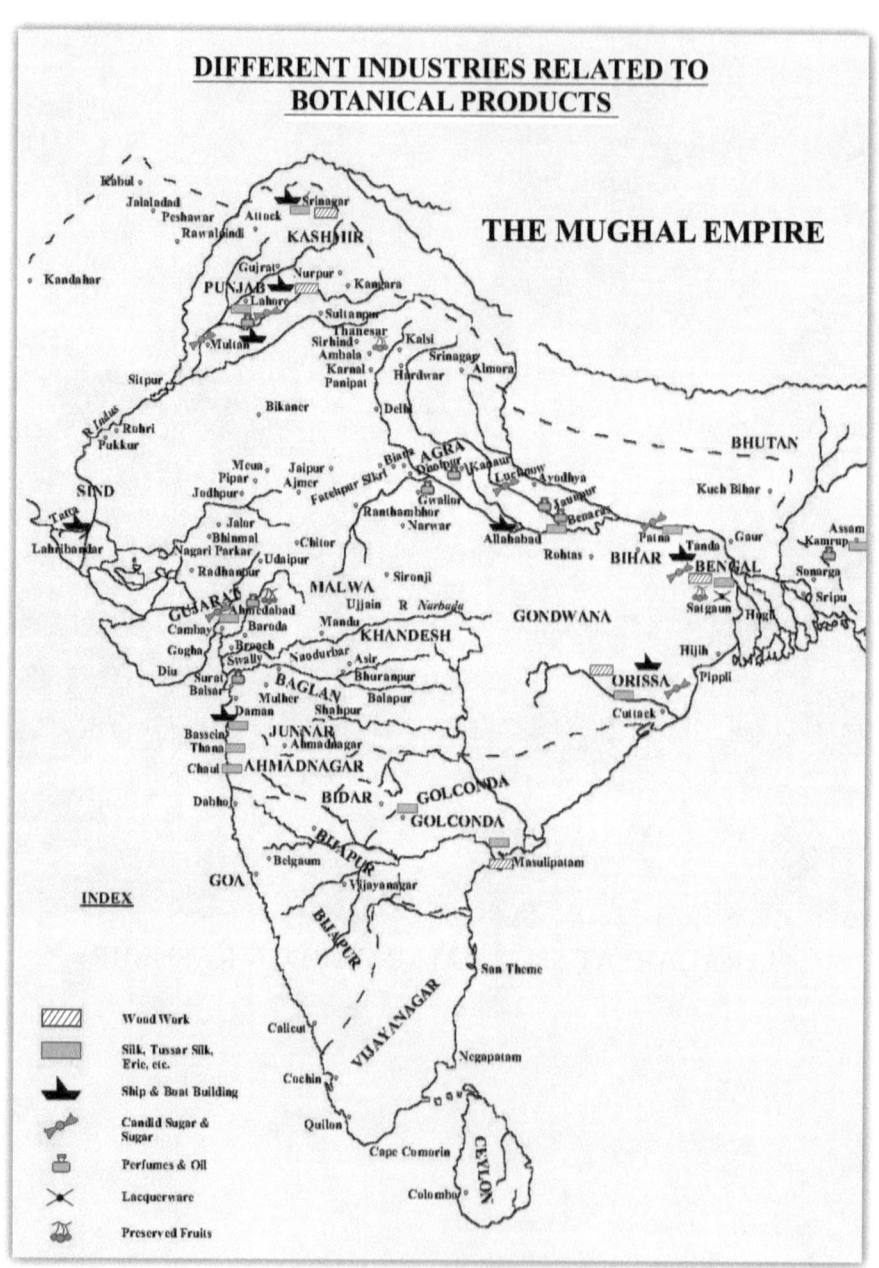

Botanical Products

CHASHMA SHAHI KASHMIR

CHASHMA SHAHI KASHMIR_2

CHINAR TREE IN KASHMIR

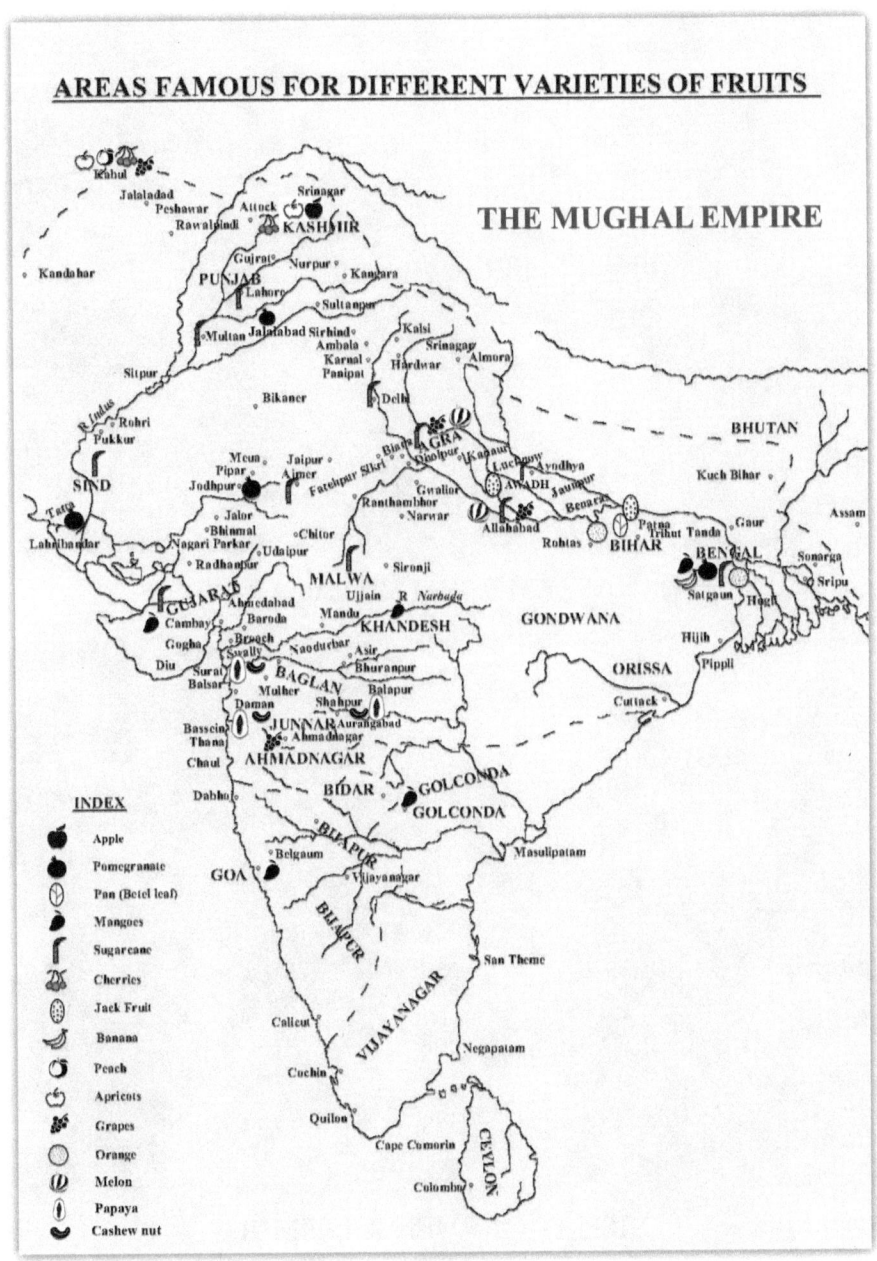

Fruits

NASEEM BAGH SRINAGAR

NISHAT GARDEN KASHMIR

pinjaur_garden

SHALAMAR KASHMIR

TERRACE OF NISHAT GARDEN KASHMIR

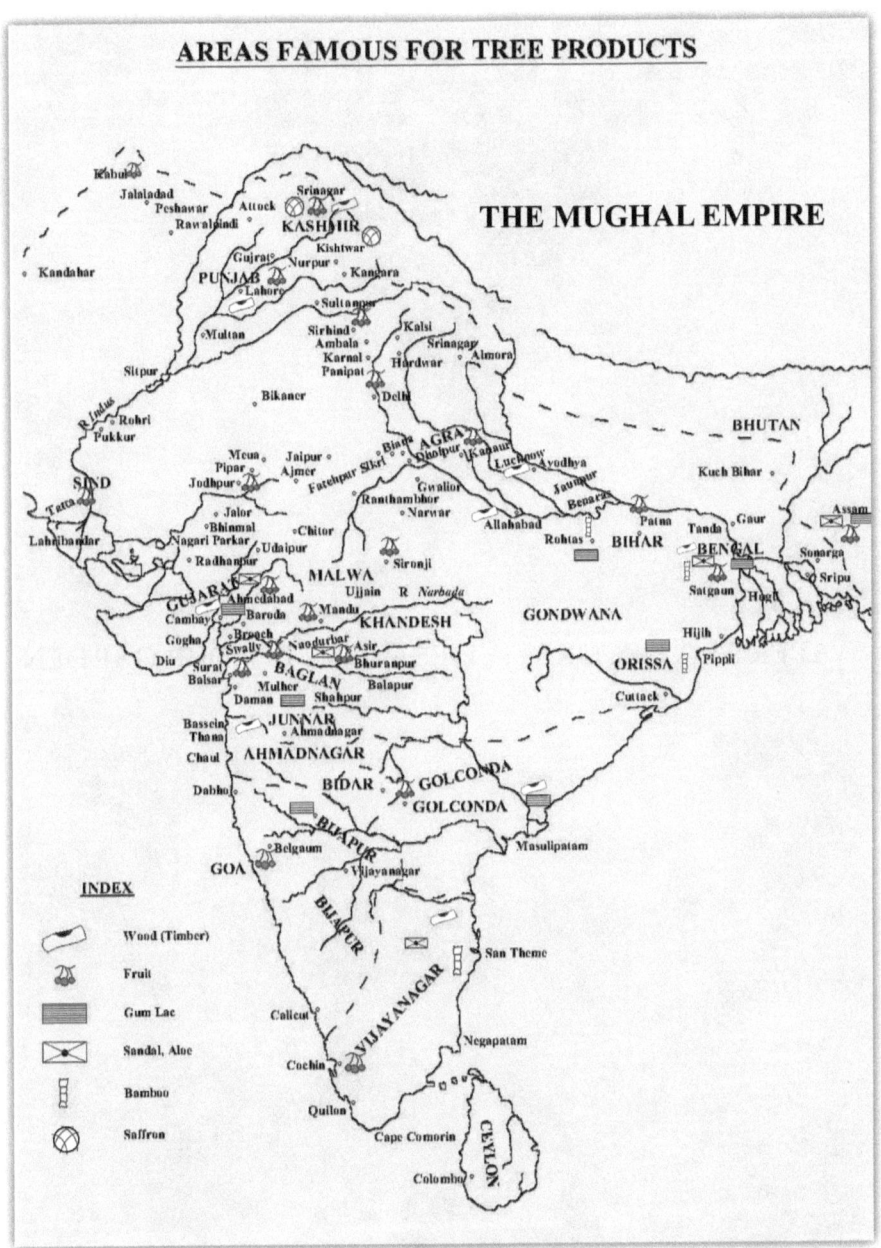

Tree products

WATER AS A DECORATIVE FEATURE PINJOUR GARDEN

WATER SYSTEM AT SHALIMAR GARDEN KASHMIR

www.ingramcontent.com/pod-product-compliance
Lightning Source LLC
Chambersburg PA
CBHW020733180526
45163CB00001B/220